# Reviews of Environmental Contamination and Toxicology

VOLUME 145

**Springer**
*New York*
*Berlin*
*Heidelberg*
*Barcelona*
*Budapest*
*Hong Kong*
*London*
*Milan*
*Paris*
*Santa Clara*
*Singapore*
*Tokyo*

# Reviews of Environmental Contamination and Toxicology

Continuation of Residue Reviews

Editor
George W. Ware



VOLUME 145

Springer

# Coordinating Board of Editors



Springer-Verlag
*New York*: 175 Fifth Avenue, New York, NY 10010, USA
*Heidelberg*: 69042 Heidelberg, Postfach 10 52 80, Germany

Library of Congress Catalog Card Number 62-18595.
Printed in the United States of America.

ISSN 0179-5953

Printed on acid-free paper.



ISBN 0-387-94584-9 Springer-Verlag New York Berlin Heidelberg

# Foreword

International concern in scientific, industrial, and governmental communities over traces of xenobiotics in foods and in both abiotic and biotic environments has justified the present triumvirate of specialized publications in this field: comprehensive reviews, rapidly published research papers and progress reports, and archival documentations. These three international publications are integrated and scheduled to provide the coherency essential for nonduplicative and current progress in a field as dynamic and complex as environmental contamination and toxicology. This series is reserved exclusively for the diversified literature on "toxic" chemicals in our food, our feeds, our homes, recreational and working surroundings, our domestic animals, our wildlife and ourselves. Tremendous efforts worldwide have been mobilized to evaluate the nature, presence, magnitude, fate, and toxicology of the chemicals loosed upon the earth. Among the sequelae of this broad new emphasis is an undeniable need for an articulated set of authoritative publications, where one can find the latest important world literature produced by these emerging areas of science together with documentation of pertinent ancillary legislation.

Research directors and legislative or administrative advisers do not have the time to scan the escalating number of technical publications that may contain articles important to current responsibility. Rather, these individuals need the background provided by detailed reviews and the assurance that the latest information is made available to them, all with minimal literature searching. Similarly, the scientist assigned or attracted to a new problem is required to glean all literature pertinent to the task, to publish new developments or important new experimental details quickly, to inform others of findings that might alter their own efforts, and eventually to publish all his/her supporting data and conclusions for archival purposes.

In the fields of environmental contamination and toxicology, the sum of these concerns and responsibilities is decisively addressed by the uniform, encompassing, and timely publication format of the Springer-Verlag (Heidelberg and New York) triumvirate:

*Reviews of Environmental Contamination and Toxicology* [Vol. 1 through 97 (1962–1986) as Residue Reviews] for detailed review articles concerned with any aspects of chemical contaminants, including pesticides, in the total environment with toxicological considerations and consequences.

*Bulletin of Environmental Contamination and Toxicology* (Vol. 1 in 1966) for rapid publication of short reports of significant advances and discoveries in the fields of air, soil, water, and food contamination and pollution as well as methodology and other disciplines concerned with the introduction, presence, and effects of toxicants in the total environment.

*Archives of Environmental Contamination and Toxicology* (Vol. 1 in 1973) for important complete articles emphasizing and describing original experimental or theoretical research work pertaining to the scientific aspects of chemical contaminants in the environment.

Manuscripts for *Reviews* and the *Archives* are in identical formats and are peer reviewed by scientists in the field for adequacy and value; manuscripts for the *Bulletin* are also reviewed, but are published by photo-offset from camera-ready copy to provide the latest results with minimum delay. The individual editors of these three publications comprise the joint Coordinating Board of Editors with referral within the Board of manuscripts submitted to one publication but deemed by major emphasis or length more suitable for one of the others.

Coordinating Board of Editors

# Preface

Worldwide, anyone keeping abreast of current events is exposed daily to multiple reports of environmental insults: global warming (greenhouse effect) in relation to atmospheric $CO_2$, nuclear and toxic waste disposal, massive marine oil spills, acid rain resulting from atmospheric $SO_2$ and $NO_x$, contamination of the marine *commons*, deforestation, radioactive contamination of urban areas by nuclear power generators, and the effect of free chlorine and chlorofluorocarbons in reduction of the earth's ozone layer. These are only the most prevalent topics. In more localized settings we are reminded of exposure to electric and magnetic fields; indoor air quality; leaking underground fuel tanks; increasing air pollution in our major cities; radon seeping from the soil into homes; movement of nitrates, nitrites, pesticides, and industrial solvents into groundwater; and contamination of our food and feed with bacterial toxins. Some of the newer additions to the vocabulary include *xenobiotic transport*, *solute transport*, *Tiers 1 and 2*, *USEPA to cabinet status*, and *zero-discharge*.

It then comes as no surprise that ours is the first generation of mankind to have become afflicted with the pervasive and acute fear of chemicals, appropriately named *chemophobia*.

There is abundant evidence, however, that virtually all organic chemicals are degraded or dissipated in our not-so-fragile environment, despite efforts by environmental ethicists and the media to persuade us otherwise. But for most scientists involved in reduction of environmental contaminants, there is indeed room for improvement in all spheres.

Environmentalism has become a global political force, resulting in multinational consortia emerging to control pollution and in the maturation of the environmental ethic. Will the new politics of the next century be a consortium of technologists and environmentalists or a progressive confrontation? These matters are of genuine concern to governmental agencies and legislative bodies around the world, for many chemical incidents have resulted from accidents and improper use.

For those who make the decisions about how our planet is managed, there is an ongoing need for continual surveillance and intelligent controls, to avoid endangering the environment, wildlife, and the public health. Ensuring safety-in-use of the many chemicals involved in our highly industrialized culture is a dynamic challenge, for the old established materials are continually being displaced by newly developed molecules more acceptable to environmentalists, federal and state regulatory agencies, and public health officials.

Adequate safety-in-use evaluations of all chemicals persistent in our air, foodstuffs, and drinking water are not simple matters, and they incorporate the judgments of many individuals highly trained in a variety of complex biological, chemical, food technological, medical, pharmacological, and toxicological disciplines.

*Reviews of Environmental Contamination and Toxicology* continues to serve as an integrating factor both in focusing attention on those matters requiring further study and in collating for variously trained readers current knowledge in specific important areas involved with chemical contaminants in the total environment. Previous volumes of *Reviews* illustrate these objectives.

Because manuscripts are published in the order in which they are received in final form, it may seem that some important aspects of analytical chemistry, bioaccumulation, biochemistry, human and animal medicine, legislation, pharmacology, physiology, regulation, and toxicology have been neglected at times. However, these apparent omissions are recognized, and pertinent manuscripts are in preparation. The field is so very large and the interests in it are so varied that the Editor and the Editorial Board earnestly solicit authors and suggestions of underrepresented topics to make this international book series yet more useful and worthwhile.

*Reviews of Environmental Contamination and Toxicology* attempts to provide concise, critical reviews of timely advances, philosophy, and significant areas of accomplished or needed endeavor in the total field of xenobiotics in any segment of the environment, as well as toxicological implications. These reviews can be either general or specific, but properly they may lie in the domains of analytical chemistry and its methodology, biochemistry, human and animal medicine, legislation, pharmacology, physiology, regulation, and toxicology. Certain affairs in food technology concerned specifically with pesticide and other food-additive problems are also appropriate subjects.

Justification for the preparation of any review for this book series is that it deals with some aspect of the many real problems arising from the presence of any foreign chemical in our surroundings. Thus, manuscripts may encompass case studies from any country. Added plant or animal pest-control chemicals or their metabolites that may persist into food and animal feeds are within this scope. Food additives (substances deliberately added to foods for flavor, odor, appearance, and preservation, as well as those inadvertently added during manufacture, packing, distribution, and storage) are also considered suitable review material. Additionally, chemical contamination in any manner of air, water, soil, or plant or animal life is within these objectives and their purview.

Normally, manuscripts are contributed by invitation, but suggested topics are welcome. Preliminary communication with the Editor is recommended before volunteered review manuscripts are submitted.

Department of Entomology
University of Arizona
Tucson, Arizona

G.W.W.

# Table of Contents

# Environmental Hazards of Aluminum to Plants, Invertebrates, Fish, and Wildlife

Donald W. Sparling* and T. Peter Lowe*

## Contents



*National Biological Service, Patuxent Environmental Science Center, Laurel, MD 20708 U.S.A.


*Reviews of Environmental Contamination and Toxicology, Vol. 145.*

## I. Introduction

Aluminum (Al) is the third most common mineral and the most common metal in Earth's crust, accounting for approximately 8.1% of the crust by weight. Thus, it cannot be considered a contaminant in the usual sense of the word. However, despite its near omnipresence throughout the world, Al has been of major concern as a primary limiting factor to cultivated plants for several decades. In much of the world, Al severely restricts the growth and presence of plant species. Since the late 1970s, concern about Al toxicity has spread to natural habitats, most notably forests and aquatic communities. The primary impetus for this concern has been the increased awareness of the effects of anthropogenic acidification through mine drainage, acid deposition, and other sources. The toxicity of Al is intimately associated with pH in that the metal is soluble and biologically available in acidic (pH $<5.5$) soils and waters but relatively innocuous in circumneutral (pH 5.5–7.5) conditions. Forest die-offs and reduced survivorship or impaired reproduction of aquatic invertebrates, fish, and amphibians have been directly connected to Al toxicity. Indirect effects on birds and mammals also have been identified. The purpose of this review is to summarize the toxic effects of Al to populations and to evaluate the potential hazards to the communities in which these populations are found.

## II. Sources and Environmental Chemistry of Aluminum

Although Al is ubiquitous in soils, it varies from 0.4% of total weight in carbonates to 9.4% in marine clays (Table 1). Across the United States, soils in Florida and parts of Georgia, Texas, Oklahoma, and Michigan are less than 2.0% Al, whereas portions of the Pacific Northwest, New England, Colorado, and Nevada have concentrations greater than 8.0%.

Aluminum can occur in many forms in soil. The metal occurs principally as tetrahedral and octahedral crystals of aluminosilicates. However, it may also combine with other substances to form amorphous or crystalline clays and sesquioxides, aluminum phosphates, or ionically bound organic compounds (Driscoll and Schecher 1990).

The biogeochemical cycling of the metal is complex (Driscoll and Schecher 1988). In the lithosphere, its primary reservoir, Al may be found as part of the mineral component of the soil or bound to charged soil and organic particles. Under circumneutral to slightly alkaline conditions (pH 6.0–8.0), Al has low solubility and is essentially biologically inactive. In alkaline soils and solutions (pH $>8.0$), the solubility of Al increases but its bioavailability is poorly known. Weathering or acidification to pH below 5.5 increases the dissolution kinetics of Al and places some of the metal into solution. Its dissolved form is most readily assimilated by living organisms.

Once in solution, Al may combine with several organic complexes, especially oxalic, humic, and fulvic acids. The metal may also combine

Table 1. Synoptic list of ambient aluminum levels in soil and water.

| Source | Aluminum level | pH | Comments | References |
|---|---|---|---|---|
| Igneous rock | 81,300 $mg \cdot kg^{-1}$ | | composite | 1 |
| Basalt | 83,000 $mg \cdot kg^{-1}$ | | composite | 1 |
| Granite | 67,000 $mg \cdot kg^{-1}$ | | composite | 1 |
| Sandstone | 25,000 $mg \cdot kg^{-1}$ | | composite | 1 |
| Limestone | 4000 $mg \cdot kg^{-1}$ | | composite | 1 |
| Shale | 82,000 $mg \cdot kg^{-1}$ | | composite | 1 |
| Clay | 86,000 $mg \cdot kg^{-1}$ | | composite | 1 |
| Marine clays | 94,000 $mg \cdot kg^{-1}$ | | composite | 2 |
| Coal | 10,000 $mg \cdot kg^{-1}$ | | | 2 |
| Atmosphere | 0.04–0.39 $\mu g \cdot m^{-3}$ | | median value | 3 |
| Precipitation | 520–1,120 $\mu g \cdot L^{-1}$ | | median value | 3 |
| Snow | 56–1,300 $\mu g \cdot L^{-1}$ | | median value | 3 |
| Adirondack lakes | 4.3–21.9 $\mu g \cdot L^{-1}$ | 5.9–7.1 | 155 lakes sampled | 4 |
| Adirondack lakes | 69.9–208 $\mu g \cdot L^{-1}$ | 4.7–5.1 | 179 acidic lakes | 4 |
| Maine lakes | 18.9–472 $\mu g \cdot L^{-1}$ | 4.7–6.8 | 5 lakes | 5 |
| Florida | Median, 37.8 $\mu g \cdot L^{-1}$ | 4.9 | 260 Panhandle lakes | 6 |
| Florida | Median, 56.7 $\mu g \cdot L^{-1}$ | 6.4 | 845 northern lakes | 6 |
| Florida | Median, 40.5 $\mu g \cdot L^{-1}$ | 6.8 | 945 southern lakes | 6 |
| Florida | 20–270 $\mu g \cdot L^{-1}$ | 4.4–5.6 | 14 acidic lakes | 6 |
| Wisconsin | 35–172 $\mu g \cdot L^{-1}$ | 4.6–5.5 | 4 clearwater lakes | 7 |
| New Jersey | 13–508 $\mu g \cdot L^{-1}$ | 3.6–6.2 | 6 acidified lakes | 8 |
| Ontario, Canada | 73–1097 $\mu g \cdot L^{-1}$ | 4.0–5.6 | 7 lakes near smelter | 9 |
| Coal pile leachings | 6–480 $\mu g \cdot L^{-1}$ | 6.6–4.8 | above and below drainage | 10 |
| Kentucky stream | 65–26,500 $\mu g \cdot L^{-1}$ | 8.5–2.6 | above and below acid mine drainage | 11 |

References: (1) Goldschmidt 1958; (2) Freedman and Hutchinson 1986; (3) Havas 1986a,b; (4) Driscoll et al. 1991; (5) Kahl et al. 1991; (6) Pollman and Canfield 1991; (7) Cook and Jager 1991; (8) Sprenger et al. 1987; (9) Yan and Dillon 1984; (10) Tan and Coler 1986; (11) Short et al. 1990.

with inorganic molecules including sulfate ($SO_4^{2-}$), fluoride ($F^-$), phosphates ($PO_3^{3-}$), bicarbonates ($HCO_3^-$), or hydroxides ($OH^-$), depending on the relative concentrations of these anions. Biological activity and toxicity vary with composition. For example, Al sulfates are generally considered less toxic than hydroxide or organically bound Al (Driscoll and Schecher 1988). Aqueous Al ($Al^{3+}$), however, is more chemically and biologically active than that bound in soil or sediments.

The biological cycling of Al is considered unimportant to the overall balance of the metal. It is most likely not an essential element to life (National Academy of Sciences 1980) and is not stored in appreciable quantities within any tissue. Most ingested Al is rapidly excreted; it then reenters reservoirs in the lithosphere or hydrosphere.

The environmental chemistry of Al is essentially driven by pH, hence its toxicity is interwoven with that of hydrogen ion ($H^+$). Increased mobilization in acidic soils leads to a series of events including (1) greater availability to plants through increased concentrations in soil solutions; (2) general downward leaching in temperate podzolic soils with decreased concentrations in upper layers (i.e., A horizon) and concomitant increases in lower layers (B horizon); and (3) gradual removal from soil into streams and lakes of acidified watersheds. The bioavailability of Al is reduced in organically rich soils due to binding between cationic forms of Al and negatively charged organic molecules. High levels of phosphate in soils may also reduce bioavailability by precipitating $AlPO_4$. The cation-exchange capacity of soils, which is largely determined by calcium ($Ca^{2+}$), magnesium ($Mg^{2+}$), potassium ($K^+$), and sodium ($Na^+$) ions, also ameliorates the availability of Al through competition for binding sites on soil particles (Cronan and Schofield 1990).

Once in solution, Al can take several forms, depending on pH and, secondarily, on other constituents. Driscoll and Schecher (1990) presented a categorization of Al species that is widely used in the scientific literature. Total Al refers to all forms, dissolved and undissolved, in water. Dissolved, or monomeric Al, can be separated into organically bound (nonlabile) and labile ($Al^{3+}$), with $Al^{3+}$ possibly combining with one of the anions mentioned previously. The concentration of the labile form increases with acidity. Aluminum concentrations in streams and lakes impacted by acidification tend to follow seasonal trends, with the greatest levels in early spring following snowmelt (Herrmann et al. 1989; Sprenger et al. 1987).

The greatest concern for Al toxicity is associated with anthropogenic acidification of watersheds. Numerous studies have found that concentrations of the more toxic monomeric species increase in soil and water with acidification. For example, in circumneutral, oligotrophic waters Al concentrations range from 0.2 to 20 $\mu g \cdot L^{-1}$ but may reach 1,000 $\mu g \cdot L^{-1}$ in some acidified lakes (Havas 1986a). However, dissolved Al levels may naturally exceed 10,000 $\mu g \cdot L^{-1}$ in acidic bogs (see Table 1).

The most extensive source of acidification in natural waterways is wet and dry acid deposition. Principal sources of this deposition include industrial and automotive emission. During the 1980s, an extensive effort was expended on determining the extent and effects of acid deposition in North America (NAPAP 1991). It was estimated that, within most of the U.S., 20% of both streams and lakes could be considered acidic or threatened with acidification. Not all of these waterways are becoming acidified by anthropogenic processes, but many streams and lakes, especially in the northeastern part of the country and southern Canada, are affected by acid deposition. Many waters with pH below 5.5 have elevated levels of Al.

Surface mining can be an important factor in Al contamination of water because of the acidity associated with mining and because deeper soil horizons with higher metal concentrations are exposed to weathering and leaching. For example, river water pH dropped from 8.5 just above an acid mine inflow to a Kentucky river to 2.6 just below the inflow. Total Al levels increased from 65 $\mu g \cdot L^{-1}$ above to 26,500 $\mu g \cdot L^{-1}$ below the inflow. At 2.4 km below the inflow, pH returned to 8.3 and total Al was 219 $\mu g \cdot L^{-1}$ (Short et al. 1990).

Leachate from a coal pile that abutted a Massachusetts stream decreased pH in the stream from 6.6 to 4.8 and increased total Al from 6 $\mu g \cdot L^{-1}$ to 480 $\mu g \cdot L^{-1}$ (Tan and Coler 1986). Concentration of the monomeric form was positively correlated to acidity among 1977 spoil banks produced by surface mining of coal in Illinois (Haynes and Klimstra 1975). Spoil banks with pH below 3.1 had Al to 1920 $mg \cdot kg^{-1}$ ($\bar{x}$ = 525 $mg \cdot kg^{-1}$), those with pH 4.1–5.0 to 480 $mg \cdot kg^{-1}$ ($\bar{x}$ = 13.6 $mg \cdot kg^{-1}$), and those with pH above 8.0, $\bar{x}$ = 2.4 $mg \cdot kg^{-1}$ Al.

Other perturbations of soil or sediments can also increase concentrations of toxic Al. Extensive logging in an experimental forest decreased the pH of a nearby stream from 5.4 to 4.6 and increased labile Al more than sixfold from 27 $\mu g \cdot L^{-1}$ to 1620 $\mu g \cdot L^{-1}$ (Lawrence et al. 1987); levels remained elevated for more than 8 mon following removal of the trees. Alum [$Al_2(SO_4)_3$] used to precipitate particulates from wastewater can increase aqueous Al when discharged into acidic streams.

## III. Effects on Plants

Aluminum has been recognized as a major limiting factor for plants in acid soils for over 50 years, and numerous original and summary papers have been written about the metal's phytotoxic effects in terrestrial and aquatic ecosystems (e.g., Crowder 1991; Fageria et al. 1988; Havas 1986b; Taylor 1988). Elevated levels in acidic soils, sediments, and water can alter the species composition of primary producers, reduce their vigor, and raise Al concentrations in wildlife foods.

### A. Aquatic Plants

It is difficult to generalize about the toxic effects of Al in aquatic plants or other organisms because comparisons among studies are confounded by differences in experimental design, water chemistry, and physical characteristics. Studies vary from whole-lake or *in situ* experiments that examine toxicity under actual conditions to more rigidly controlled mesocosm or laboratory toxicity tests. Field studies are particularly difficult to compare because of differences in pH and concentrations of organic ligands or other cations. Controlled mesocosm studies present a more accurate analysis of the direct toxic effects of Al, but their simplified environments may produce results that cannot be directly applied to ponds and lakes. The most reliable information comes from a combination of both types of sources, but such studies are rare.

Aquatic plants apparently can tolerate higher levels of Al and acidity than aquatic invertebrates, amphibians, or fish. For example, several species of algae have been reported at pH less than 3.0 and dissolved Al above 100 mg·L$^{-1}$ (Table 2) (Havas 1986b). *Charcium* spp., *Euglena mutabilis*, and *Pinnularia acoricola* survive at Al levels of 2500 mg·L$^{-1}$ or greater.

The species compositions of algae typically change with reduced pH or elevated Al. Often, dominant species change and overall species richness decreases. For example, Havens and DeCosta (1987) compared algal communities in polyethylene mesocosms filled with lake water under three treatments: acidification to pH 4.5 (Ac), acidification with 300 $\mu$g·L$^{-1}$ total Al (Ac + Al), and control (pH 7.0). Algal responses in both the Ac and Ac + Al treatments were similar. *Euglena* spp., *Chlorella vulgaris, Dinobryon divergens*, and *Arachnochloris minor* diminished or disappeared in both treatments as pH decreased, and dissolved Al increased to 180 $\mu$g·L$^{-1}$. In contrast, *Chlamydomonas mucicola* and *Peridinum inconspicuum* increased in density with declining pH. The authors concluded that acid-tolerant species of algae were also likely to be Al tolerant.

In general, overall algal density declines and diatom and chrysophyte communities are replaced by dinoflagellates and chlamydomonads with elevated Al. Filamentous green algae, especially *Mougoetia*, however, may increase in biomass (Havens and Heath 1990). Under a regimen of slow acidification of mesocosms (pH drop from 8.5 to 4.5 and Al increase from

Table 2. Algae species that can live at aqueous Al levels greater than 1000 $\mu$g·L$^{-1}$.[a]

| Species | Al level ($\mu$g·L$^{-1}$) |
|---|---|
| *Chlamydomonas acidophila* | 810 |
| *Chlamydomonas applanata* | 1570 |
| *Characium* sp. | 2500 |
| *Cryptomonas* sp. | 600 |
| *Euglena mutabilis* | 3130 |
| *Eunotia arcus* | 212 |
| *Eunotia glacialis* | 800 |
| *Hormidium rivulare* | 411 |
| *Lepocinclis ovum* | 400 |
| *Migrathomnion strictissium* | 169 |
| *Nitschia communis* | 800 |
| *Nitschia elliptica* | 411 |
| *Nitschia subcapitellata* | 325 |
| *Pinnularia acoricola* | 2500 |
| *Ulothrix zonata* | 138 |
| *Zygongonium ericetorum* | 411 |

[a]After Havas 1986a,b.

below detection limits to 180 $\mu g \cdot L^{-1}$), filamentous blue-green algae, particularly *Aphanizomenon flos-aquae*, was replaced by cryptophytes, chrysophytes, and nanochlorophytes. *Peridinium inconspicuum*, *Cryptomonas erosa*, and *Chlamydomonas globosa* predominated early during acidification but subsequently were replaced by *Closterium* spp. The addition of Al did not alter the final outcome of the experiment but caused some of the changes to occur earlier than acidification alone.

Pillsbury and Kingston (1990) demonstrated an effect of Al beyond that of acidification alone. Their treatments consisted of lake water (pH 5.7) with 50, 100, or 200 $\mu g \cdot L^{-1}$ Al, untreated lake water (pH 5.7), and water acidified to pH 4.7 without Al addition. The diatom *Asterionella ralfsii* dominated in the lake water and acid treatment but decreased with the addition of 50 $\mu g \cdot L^{-1}$ Al and nearly disappeared at 100 $\mu g \cdot L^{-1}$; various desmid species showed similar patterns. Population levels of *Peridinum limbarum, Dinobryon bavaricum,* and *Elaklothrix* spp. declined with the addition of 200 $\mu g \cdot L^{-1}$ Al.

Aluminum may bind with phosphorus (P) and reduce its availability to primary producers. Nalewajko and Paul (1985) observed a greater precipitation of both elements at pH 6.1 and 6.9 than at pH 4.5 and that precipitation of P increased with increasing levels of Al. Phosphorus uptake by phytoplankton was suppressed at the higher pHs with the addition of Al. Additional P resulted in greater uptake of P at pH 6.9 but not at 6.1 or 4.5. At all three pH levels, the addition of Al decreased photosynthetic rates of algae.

Elevated $H^+$, reduced Ca, and reduced P may interact with Al under acidic conditions to influence plant growth. As a result, it is difficult to discriminate between the direct toxic effects of Al and those of factors associated with reduced pH. For example, Jackson and Charles (1988) investigated relationships between presence and absence of floating-leaved and submersed macrophytes and water chemistry in 29 lakes in the Adirondack mountains. The pH of these lakes ranged from 4.5 to 7.8 and Al levels from 5 to 673 $\mu g \cdot L^{-1}$. Total Al, soluble Al, alkalinity, Ca, Mg, Na, conductivity, and elevation were important in predicting the presence or absence of plant species. However, these factors were also highly correlated with pH. A gradient in species composition corresponded to pH, and the authors concluded that acidity was the principal factor limiting plant distribution, whereas Al was only secondary.

Ormerod et al. (1987a) found that Al levels in aquatic plants correlated inversely with pH among 85 stream sampling sites in Wales. The range in mean pH among these sites was 4.9–6.2 and that of mean filterable Al was 48–227 $\mu g \cdot L^{-1}$. Acidity, followed by $Al^{3+}$, correlated negatively to the presence or absence of aquatic macrophytes. When pH was removed from the analysis, Al became the dominant water-quality characteristic. In contrast, Catling et al. (1986) found that Al was not important in determining species distributions of aquatic macrophytes in 20 lakes of Kejimkujik National Park, Nova Scotia. Levels ranged from 35 to 250 $\mu g \cdot L^{-1}$ for Al and 4.4 to 6.0 for pH.

Because aqueous Al is frequently negatively correlated with pH (Sprenger and McIntosh 1989), many plants that grow in acidic conditions are probably Al tolerant. This generalization is supported by studies on algae reported earlier. Therefore, certain species such as *Eriocaulon septangulare, Lobelia dortmana, Eleocharis acicularis, Sphagnum* spp., and *Scirpus subterminalis* (Table 3) may be considered Al tolerant. A few species that grow well under acidic conditions (e.g., *Potamogeton confervoides, Nuphar microphyllum, Isoetes acadiensis, I. macrospora*) (Catling et al. 1986) actually may be stimulated by Al. Alternatively, these species may simply be tolerant and better able to compete with sensitive species under acidic conditions.

## B. Forest Species

Declines in forest vigor and die-backs of some stands have been observed in Germany, central Europe, Scandinavia, U.S., and Canada during the past 30 years (Krahl-Urban et al. 1990). Over 50% of the forests in Germany have declined, particularly the southern fir forests. Beech, oak, pine, and spruce forests have also been affected. In North America, die-backs or reduced growth are obvious at high elevation in red spruce (*Picea rubens*) of the Appalachian mountains, balsam (*Abies balsamea*) and Fraser firs (*A. fraseri*) in New England and Canada, loblolly (*Pinus taeda*) and slash pine (*P. elliottii*) in commercial forests of the southeast, and sugar maples (*Acer saccharum*) in eastern U.S. and southeastern Canada. In spruce–fir forests, signs of these declines include reduced growth of trees, needle discoloration and loss, crown thinning, apical bud die-back, and branch and tree mortality; 26%–37% of some stands may consist of dead trees (Adams and Eagar 1992).

Several species of conifers grown in Al-enriched solutions in laboratories experience reduced root growth rates. For example, growth significantly declined at pH 3.8 in red and white spruce (*Picea glauba*) at 50 $mg \cdot L^{-1}$ Al, jack pine (*Pinus banksiana*) at 40 $mg \cdot L^{-1}$, and white pine (*Pinus strobus*) at 80 $mg \cdot L^{-1}$ (Hutchinson et al. 1986). At 1.3 $mg \cdot L^{-1}$ Al and pH range 3.2–4.6, white spruce seedlings had shorter roots, less root mass, and lower root : shoot ratios than controls (Nosko et al. 1988). In European birch (*Betula pendula*) and Norway spruce (*Picea abies*), 2.7 $mg \cdot L^{-1}$ Al reduced root elongation, whereas 13.5 $mg \cdot L^{-1}$ Al was required before Scotch pine (*Pinus sylvestris*) showed effects (Eldhuset et al. 1987). Other toxic levels of Al included 3 $mg \cdot L^{-1}$ (pH 3.9) in *Populus* hybrids (Steiner et al. 1984), 3 $mg \cdot L^{-1}$ (pH unreported) in peach (*Prunus persica*) (Baes and McLaughlin 1987), and 27 $mg \cdot L^{-1}$ in sugar maple (Thornton et al. 1986). When exposed to 2.7 $mg \cdot L^{-1}$ Al (pH 4.2–5.4) for 8 wk, dry mass of beech (*Fagus sylvatica*) leaves, roots, and stems were 21%–44% lower than controls (Bengtsson et al. 1988). At pH 3.8, shoot growth of balsam fir was reduced by 24% when

Table 3. Partial list of aquatic plants that are presumably enhanced by, tolerant of, or highly sensitive to acidity and elevated levels of aluminum.[a]

| Enhanced species[b] | Tolerant species[c] | Sensitive species[d] |
|---|---|---|
| *Eleocharis smallii* | *Brassenia schreiberi* | *Arthrodesmus indentatus* |
| *Isoetes acadiensis* | *Drepanocladus fluitans* | *Arthrodesmus octocornus* |
| *Isoetes macrospora* | *Drepanocladus exannulatus* | *Arthrodesmus quiriferus* |
| *Nuphar microphyllum* | *Elatine minima* | *Asterionella falfsii* |
| *Potamogeton confervoides* | *Eleocharis acicularis* | *Ceratophyllum demersum* |
| *Scirpus subterminalis* | *Eleocharis robbinsii* | *Chara vulgaris* |
| | *Eriocaulon septangulare* | *Elodea canadensis* |
| | *Fontinalis* spp. | *Elodea nuttallii* |
| | *Gratiola aurea* | *Elodea occidentalis* |
| | *Hyscomium armoricanum* | *Isoetes tuckermanni* |
| | *Isoetes muricata* | *Lemna minor* |
| | *Juncus peliocarpus* | *Myriophyllum verticillatum* |
| | *Lobelia dortmanna* | *Potamogeton amplifolius* |
| | *Mougeotia* sp. | *Potamogeton epihydrus* |
| | *Myriophyllum farwellii* | *Potamogeton gramineus* |
| | *M. tenellum* | *Potamogeton perfoliatus* |
| | *Najas flexilis* | *Potamogeton praelongus* |
| | *Nardia compressa* | *Potamogeton robbinsii* |
| | *Nuphar variegatum* | *Potamogeton spirillus* |
| | *Nuphar luteum* | *Sparganium fluctuans* |
| | *Nymphoides cordata* | *Spirodela polyrhiza* |
| | *Oedogonium* sp. | *Staurastrum arachnae* |
| | *Peridinum limbatum* | *Staurastrum longipes* |
| | *Pontederia cordata* | *Staurastrum pentacerum* |
| | *Rhacomitrium aciculare* | *Utricularia gibba* |
| | *Sagittaria angustifolium* | *Utricularia intermedia* |
| | *Sagittaria graminea* | *Vallisneria americana* |
| | *Scapania undulata* | |
| | *Sphagnum* sp. | |
| | *Utricularia gemniscapa* | |
| | *Utricularia purpurea* | |
| | *Utricularia resupenata* | |
| | *Utricularia vulgaris* | |

[a]After Wile and Miller 1983; Roberts et al. 1985; Catling et al. 1986; Ormerod et al. 1987b; Jackson and Charles 1988; Pillsbury and Kingston 1990.

[b]Enhanced species were found only in low-pH waters with presumably elevated levels of Al (see text) or had growth increased with Al additions.

[c]Tolerant species are found in waters with pH 4.9 or lower, where Al levels are presumably higher than in circumneutral waters (see text).

[d]Sensitive species were only found in waters with pH above 6.8, where Al is reputedly nontoxic.

seedlings were kept in a nutrient solution containing 200 $mg \cdot L^{-1}$ Al at pH 3.8 for 32 d, and 50 $mg \cdot L^{-1}$ significantly reduced root growth of both balsam fir and red spruce (Schier 1985).

Studies have shown that very low levels of Al may enhance growth, although the mechanism is unknown. For example, red spruce, balsam fir (Schier 1985), and sugar maples (Thornton et al. 1986) appear to increase shoot growth when exposed to the metal compared to plants grown in solutions with no Al.

Aluminum appears to alter the uptake of several ions by tree roots. When exposed to 2.7 $mg \cdot L^{-1}$, beech seedlings experienced an 80% decline in Ca and Mg in all plant parts compared to controls, and at 27 $mg \cdot L^{-1}$ Al, Ca dropped 90% (Bengtsson et al. 1988). Rengel (1992) suggested that Ca : Al molar ratios were important in determining Al toxicity in that ratios less than 1 were likely to be more toxic than higher ratios.

Phosphorus uptake increased at lower exposure levels but decreased when plants were exposed to 27 $mg \cdot L^{-1}$ Al. Potassium uptake increased at both levels. In the same experiment, seedlings exposed to a soil solution from the A horizon of an acidic forest soil with 5 $mg \cdot L^{-1}$ total Al, 0.5 $mg \cdot L^{-1}$ labile Al, and pH 3.7 did not differ from controls in ion uptake. However, when exposed to a solution taken from the B horizon with a labile Al concentration of 11 $mg \cdot L^{-1}$ and similar pH, seedlings lost Ca and Mg and gained P at a rate similar to that of 27 $mg \cdot L^{-1}$ Al in solution.

Probably by competing for binding sites, Al can decrease the uptake of Cd and other heavy metals under acidic conditions. A solution with 2.7 $mg \cdot L^{-1}$ Al decreased Cd uptake of *Pinus abies* by 40%, and 13 $mg \cdot L^{-1}$ Al depressed Cd uptake by 86%. Cadmium uptake could be further depressed by increasing Ca levels. Aluminum depressed concentrations of several minerals in the needles and roots of red spruce and balsam fir (Schier 1985). For spruce roots, the ranking of least to most affected minerals was $P < K < Cu < Mg < Ca = Zn = Fe < Mn$, whereas the order in fir was $P < Cu = Zn < K = Ca = Mn < Mg < Fe$.

## C. Cultivated Plants

Aluminum is widely recognized as one of the most important limiting factors to crop production in soils with a pH below 5.5 (Taylor 1988). A casual examination of bibliographic compilation sources shows that one to seven new scientific papers on Al toxicity in crop plants appear each week. Many reviews (e.g., Fageria et al. 1988; Foy 1988; Parker et al. 1989; Taylor 1988) have appeared. A complete review of this literature is beyond this paper, but some of this extensive bank may be useful in understanding Al toxicity to plants.

High levels are particularly a problem in tropical soils (e.g., oxisols, utisols, and inceptisols) (Fageria et al. 1988), and up to 50% of the world's nonirrigated croplands or 40% of all arable lands may be affected negatively (Taylor 1988). Approximately 1 billion ha in the tropics and 470 million ha in temperate regions are susceptible (Fageria et al. 1988).

Several species of crop plants have cultivars or ecotypes that differ in their tolerance to Al phytotoxicity. Among these are rice (*Oryza sativa*), corn (*Zea mays*), wheat (*Triticum aestivum*), oats (*Avena sativa*), kaffir (*Sorghum bicolor*), sugarcane (*Saccharum officinarium*), rye grass (*Lolium* spp.), canary grass (*Phalaris aquatica*), rye (*Secale cereale*), bluestem grass (*Bothriochloa* spp.), barley (*Hordeum vulgare*), amaranthus (*Amaranthus* spp.), lima beans (*Phaseolus vulgaris*), soybean (*Glycine max*), string beans (*Centrosema* spp.), lespedeza (*Lespedeza cuneata*), alfalfa (*Medicago sativa*), tomato (*Lycopersicon esculentum*), tobacco (*Nicotiana tabacum*), sweet potato (*Ipomoea batatas*), and sunflower (*Helianthus annuus*) (Fageria et al. 1988; Foy 1988).

Taylor (1988) summarized possible mechanisms for differences among crop plant sensitivities. Some of these mechanisms may also pertain to forest species, but there have not been sufficient studies to make generalizations. A brief synopsis of each mechanism is provided here:

1. Differences in the ability to assimilate Al: Some tolerant cultivars of wheat tend to transport less Al across cell membranes than sensitive varieties.
2. Detoxification of Al after assimilation: Tolerant varieties of tea (*Camellia sinensis*) and other species produce relatively large concentrations of organic compounds such as oxalate and citrate, which complex with Al and reduce its toxicity.
3. High rate of root growth: Because Al tends to inhibit root elongation, cultivars that have greater rates of root growth may overcome some of the toxic effects.
4. Higher cellular respiration: Greater rates of respiration may alter the uptake of cations, including Al, and make them less toxic.
5. Local enhancement of pH: The roots of some cultivars of corn, barley, wheat, and rice release buffers that increase local pH and may reduce the solubility of Al. There is some speculation as to whether this mechanism occurs only in monocotyledons or in dicotyledons as well.
6. Reduced protoplasm viscosity: Aluminum tends to increase the viscosity of protoplasm and reduce overall permeability of salts. Cultivars with thinner cytoplasm may be less inhibited.
7. Partitioning of Al: Cultivars can be separated into one of three groups based on their tendency to partition Al into roots or shoots. One group (represented by wheat, barley, soybean, and snap bean) tends to exclude it from entering roots. Another group (e.g., azalea [*Azalea* sp.], cranberry [*Vaccinium oxycoccus*], rice, and triticale) concentrates the metal in roots and transports little of it to shoots. A third group (including tea, some pines, shadscale [*Atriplex hastata*], and mangrove [*Rhizophora* sp.]) selectively partitions Al into shoots.

8. Increased efficiency of P uptake or metabolism: Aluminum commonly binds with P either in the soil or in cells and makes it unavailable to plants. Some cultivars of wheat, tomato, corn, pea, and bean either assimilate P more rapidly or have greater phosphatase activity in roots.
9. Increased efficiency of Ca uptake or metabolism: The metal may also interfere with Ca uptake, and those species or cultivars that are more efficient in assimilating Ca may overcome some of the phytotoxicity.
10. Enhanced ability to handle $NO_3^-$ and ammonium ($NH_4^+$): In highly acidic soils, nitrification is inhibited and $NH_4^+$ becomes the primary source of N. Species that tolerate elevated levels of $NH_4^+$ also tend to tolerate high Al levels. Some of these species include cranberries, sugar cane, birch (*Betula verucosa*), blueberries (*Vaccinium myrtilloides*), and some grasses (*Deschampsia flexuosa, Lolium rigidum, Paspalum notatum*).
11. Increased efficiency of water uptake: Reduced root growth can lead to increased dehydration of a plant as its shoots continue to grow. Plants with more efficient osmotic systems can circumvent this negative effect.

Some crops grow more rapidly in low Al solutions than when no Al is available. Blamey et al. (1990a,b), for example, determined that the tolerant species of birdsfoot trefoil (*Lolium pedunculatus*) grew 30%–40% more rapidly with a nutrient solution containing 270 $\mu g \cdot L^{-1}$ at a pH of 4.7 than with no Al. In contrast, growth of the sensitive species of birdsfoot trefoil (*L. corniculatus*) decreased by 30%–40% with Al. At least part of the difference between the two species was due to a lower cation-exchange capacity in *L. pedunculatus*. Other species that show enhanced growth with Al include rice (3 $\mu g \cdot L^{-1}$), *Eucalyptus* spp. (1 $\mu g \cdot L^{-1}$), tea (27 $\mu g \cdot L^{-1}$), peach (17.5 $\mu g \cdot L^{-1}$), sugar beet (*Beta vulgaris*) (1 $\mu g \cdot L^{-1}$), corn (3.5 $\mu g \cdot L^{-1}$), and wheat (3 $\mu g \cdot L^{-1}$).

Monomeric and hydrolyzed forms of Al [$Al^{3+}$, $Al(OH)^{2+}$, $Al(OH)_2{}^+$, $Al(OH)_3$] are typically the most toxic, whereas polymeric and organically bound forms have slight to no phytotoxicity (Fageria et al. 1988; Taylor 1988). Often, the sum of the concentrations of monomeric Al is used to estimate the phytotoxicity of a growing medium. Although Parker et al. (1989) contended that polymeric Al can be as toxic as $Al^{3+}$ in nutrient solutions, polymeric Al is generally not soluble in soil and therefore should not be as toxic. In soil, the concentration of $Al^{3+}$ may suffice to predict toxicity. For most field work, utisols, oxisols, and podzols with a pH less than 5.5 can be suspected of having Al-toxic properties, and these soils with pH below 5.0 should be regarded as being phytotoxic to some species.

## D. Accumulation in Plant Tissues

Concentrations of Al in plant tissues are of concern because the diets of many species of animals are composed substantially of plants. Thus, food plants are potentially important exposure routes of toxicity.

Concentrations in plant tissues vary considerably depending on ambient levels of the metal, pH of the growing medium, presence or absence of complexing agents, species of plant, and portion of plant examined. In general, the low solubility in soil or water between pH 5.5 and 7.9 reduces its availability in that range, and most attention has been paid to acidic (pH $<5.0$) environments. Some organic molecules such as oxalate and citrate may increase Al phytotoxicity by enhancing its transport across the cell membrane and increasing its concentration in cell tissues.

Plant species differ considerably in their ability to accumulate Al. Haridasan (1982) described "accumulators" as those species that concentrate Al at levels greater than 1000 $mg \cdot kg^{-1}$ dry weight and "hyperaccumulators" as those that contain levels above 5000 $mg \cdot kg^{-1}$. Among nonvascular aquatic plants, several species of algae and bryophytes are accumulators or hyperaccumulators (Table 4). Miles (unpublished data) found that the alga *Enteromorpha* spp. near industrial sites in Chesapeake Bay had a mean Al level of 22,800 $mg \cdot kg^{-1}$ dry weight with a high of 35,100 $mg \cdot kg^{-1}$. Havas (1986b) discovered that algae frequently contain more than 2000 $mg \cdot kg^{-1}$ and may bioaccumulate the metal. Aquatic mosses such as *Sphagnum* spp., *Fontinalis* spp., and *Drepanocladus* spp. typically have more Al than vascular aquatic plants in the same habitats (Crowder 1991). The highest levels in bryophytes exceed 20,000 $mg \cdot kg^{-1}$ dry weight.

In general, the importance of aqueous Al levels on tissue concentrations in vascular plants should follow the gradient: submerged nonrooted > submerged rooted ≥ floating-leaved > emergent species. This is because submerged, nonrooted species are most dependent on the water column for nutrients and minerals, whereas the other categories of plants can receive some or most of their minerals from sediments (Crowder 1991). Lehtonen (1989) determined that Al levels in *Nuphar lutea*, a floating-leaved plant, correlated more highly with ambient aqueous Al than did levels in *Phragmites australis*, an emergent species. Sprenger and McIntosh (1989) also found that the correlation between Al in water and in plant tissues was higher for submerged plants than for emergents or floating-leaved species. Hyperaccumulators among aquatic, vascular macrophytes include some *Potamogeton* species, *Myriophyllum fauvellii, Eriocaulon septangulare, Eleocharis acicularis,* and some members of the genera *Isoetes* and *Juncus* (Table 4).

Sparling and Lowe (unpublished data) compared the Al levels in four species of aquatic plants growing in experimentally acidified mesocosms. The rooted, emergent species of *Polygonum spicatum* ($\bar{x}$ = 231 ± 153 $mg \cdot kg^{-1}$ Al, dry weight) and *Sparganium americanum* ($\bar{x}$ = 106 ± 60 $mg \cdot kg^{-1}$) had lower Al concentrations than the rooted, submergent *Potomogeton* sp. ($\bar{x}$ = 686 ± 407 $mg \cdot kg^{-1}$). *Potomogeton* sp. in turn had lower Al than the nonrooted submerged *Utricularia vulgaris* ($\bar{x}$ = 1246 ± 900 $mg \cdot kg^{-1}$).

Aluminum concentrations can also vary due to season and distance from source of contamination. For example, Nuorteva (1990) reported that levels

Table 4. Aluminum levels in aquatic and terrestrial plants.

| Species | Habitat | pH | Al in habitat ($mg \cdot kg^{-1}$) | Tissue | Al in plant ($mg \cdot kg^{-1}$) | Reference |
|---|---|---|---|---|---|---|
| | | I. Aquatic | | | | |
| | | A. Algae | | | | |
| *Microcystis* sp. | FW | | | entire | 620 | 1 |
| *Euglena* sp. | FW | | | entire | 1200 | 1 |
| *Spirogyra* sp. | FW | | | entire | 2800 | 1 |
| *Chara* sp. | FW | | | entire | 2800 | 1 |
| *Pithophora* sp. | FW | | | entire | 4100 | 1 |
| *Rhizoclonium* sp. | | | | entire | 4500 | 1 |
| *Enteromorpha* sp. | Chesapeake Bay | | | entire | 22,833 | 2 |
| | San Francisco Bay | | | entire | 16,025 | 2 |
| Phytoplankton | FW | | 0.19[a] | entire | 2500 | 1 |
| Phytoplankton | FW | | 0.02[a] | entire | 3700 | 1 |
| Phytoplankton | M | | | entire | 38–440 | 1 |
| Brown algae | M | | | entire | 38–440 | 1 |
| Red algae | M | | | entire | 1100 | 1 |
| | | B. Submerged, nonrooted | | | | |
| *Utricularia purpurea* | FW | 3.6–5.8 | 12,000–20,000[b] | shoots | 3590–6580 | 3 |
| *Sphagnum fuscum* | FW | | | shoots | 517 | 1 |
| *Sphagnum megallanicum* | FW | | | shoots | 250 | 1 |
| *Sphagnum nemoreum* | FW | | | shoots | 390 | 1 |
| *Sphagnum wulfianum* | FW | | | shoots | 1000 | 1 |
| *Sphagnum pylaesii* | FW | 4.9 | 42.5[c] | shoots | 32,000 | 1 |
| *Sphagnum* sp. | FW | | 2,600–7,200[b] | shoots | 2900–22,000 | 4 |
| *Fontinalis* sp. | FW | | 2,600–7,300[a] | shoots | 1500–1800 | 4 |
| *Fontinalis antipyretica* | | | | shoots | 700 | 5 |

| | | | | | | |
|---|---|---|---|---|---|---|
| | | C. Submerged, rooted | | | | |
| *Isoetes* sp. | FW | 3.6–5.8 | 12,000–20,000[b] | shoots | 400–5340 | 3 |
| *Potamogeton* sp. | FW | 3.6–5.8 | 12,000–20,000[b] | shoots | 2670–6480 | 3 |
| *Myriophyllum farwellii* | FW | 3.6–5.8 | 12,000–20,000[b] | shoots | 2850–3590 | 3 |
| *Eriocaulon* sp. | FW | 3.6–5.8 | 12,000–20,000[b] | shoots | 1350 | 3 |
| *Eriocaulon septangulare* | FW | 4.4 | 7300[a] | roots | 4100 | 4 |
| | | | 7300[a] | shoots | 3800 | |
| | | 6.5 | 7300[a] | roots | 2900 | |
| | | | 7300[a] | shoots | 2000 | |
| *Potomogeton praelongus* | FW | | | shoots | 250 | 1 |
| *Ceratophyllum demersum* | FW | | | | 785 | 1 |
| | | D. Emergent | | | | |
| *Juncus* sp. | FW | 3.6–5.8 | 12,000–20,000[b] | shoots | 111–2190 | 3 |
| *Scirpus* sp. | FW | 3.6–5.8 | 12,000–20,000[b] | shoots | 172–514 | 3 |
| *Phragmites australis* | Finnish lakes | 4.7–5.0 | 420–430[a] | leaves | b.d.–8.8 | 6 |
| | | | 8400–13,000[b] | | | |
| | | | 8400–13,000[b] | stems | b.d. | 6 |
| | | | 8400–13,000[b] | rhizome | b.d.–116 | 6 |
| *Phragmites australis* | Finnish lakes | 6.0–6.7 | 120–320[a] | leaves | b.d. | 6 |
| | | | 9400–11,000[b] | | | |
| | | | 9400–11,000[b] | stems | n.d. | 14 |
| | | | 9400–11,000[b] | rhizome | <6.0–10 | 14 |
| *Eleocharis acicularis* | FW | | 2600–7300[a] | stems | 600–7800 | 4 |

*(continued)*

Table 4. (*Continued*)

| Species | Habitat | pH | Al in habitat ($mg \cdot kg^{-1}$) | Tissue | Al in plant ($mg \cdot kg^{-1}$) | Reference |
|---|---|---|---|---|---|---|
| | | | E. Floating-leaved | | | |
| *Nuphar advena* | FW | | | petiole | 230–300 | 1 |
| | | | | leaf | 390–510 | 1 |
| | | | | flower | 270 | 1 |
| *Nuphar lutea* | Finnish lakes | 4.7–5.0 | 420–430[a] | leaves | 84–290 | 6 |
| | | | 8400–13,000[b] | petiole | 86–150 | 6 |
| | | | 8400–13,000[b] | rhizome | 31–70 | 6 |
| | | | 8400–13,000[b] | roots | 220–770 | 6 |
| | Finnish lakes | 5.9–6.7 | 120–320[a] | leaves | 13–33 | 6 |
| | | | 9400–11,000[b] | petiole | 21–38 | 6 |
| | | | 9400–11,000[b] | rhizome | 24–40 | 6 |
| | | | 9400–11,000[b] | roots | 230–900 | 6 |
| *Nymphaea odorata* | FW | | | petiole | 180–420 | 1 |
| *Nymphaea* sp. | FW | 3.6–5.8 | 12,000–20,000[b] | shoots | 87–264 | 3 |
| *Brasenia* sp. | FW | 3.6–5.8 | 12,000–20,000[b] | shoots | 395 | 3 |
| *Orontium* sp. | FW | 3.6–5.8 | 12,000–20,000[b] | shoots | 450 | 3 |
| *Pontederia* sp. | FW | 3.6–5.8 | 12,000–20,000[b] | shoots | 18 | 3 |
| | | | II. Terrestrial | | | |
| | | | A. Noncultivated | | | |
| *Ramalina stenospora* (lichen) | Louisiana | | | entire | 900–4500 | 7 |
| *Fagus sylvatica* | forest | 2.8–4.0 | | buds | 28.0 | 8 |
| | | 2.8–4.0 | | nuts | 8.4 | 8 |

| | | | | | | |
|---|---|---|---|---|---|---|
| *Oxalis acetosella* | forest | 2.8–4.0 | | rhizome | 132 | 8 |
| | | 2.8–4.0 | | leaves | 226 | 8 |
| *Carex pilulifera* | forest | 2.8–4.0 | | leaves | 157 | 8 |
| *Abies balsamea* | NS | 3.4–3.8 | 0 | needles | 19 | 9 |
| | | | 100 | needles | 252 | 9 |
| | | | 200 | needles | 296 | 9 |
| | | | 0 | roots | 164 | 9 |
| | | | 100 | roots | 1972 | 9 |
| | | | 200 | roots | 1871 | 9 |
| *Picea rubens* | SC | 3.8 | 0 | shoots | 29 | 10 |
| | | | 160 | shoots | 772 | 10 |
| | | | 0 | roots | 243 | 10 |
| | | | 160 | roots | 5351 | 10 |
| | NS | 3.4–3.8 | 0 | needles | 18 | 9 |
| | | | 100 | needles | 139 | 9 |
| | | | 200 | needles | 176 | 9 |
| | | | 0 | roots | 237 | 9 |
| | | | 100 | roots | 2456 | 9 |
| | | | 200 | roots | 3631 | 9 |
| *Picea glauca* | SC | 3.8 | 0 | shoots | 41 | 10 |
| | | | 160 | shoots | 237 | 10 |
| | | | 0 | roots | 389 | 10 |
| | | | 160 | roots | 2588 | 10 |
| *Picea mariana* | SC | 3.8 | 0 | shoots | 40 | 10 |
| | | | 160 | shoots | 704 | 10 |
| | | | 0 | roots | 335 | 10 |
| | | | 160 | roots | 3529 | 10 |
| *Picea abies* | Finnish forests | | | needles | 60–180 | 11 |

(continued)

Table 4. (*Continued*)

| Species | Habitat | pH | Al in habitat ($mg \cdot kg^{-1}$) | Tissue | Al in plant ($mg \cdot kg^{-1}$) | Reference |
|---|---|---|---|---|---|---|
| *Pinus banksiana* | SC | 3.8 | 0 | shoots | 39 | 10 |
| | | | 160 | shoots | 988 | 10 |
| | | | 0 | roots | 211 | 10 |
| | | | 160 | roots | 3654 | 10 |
| *Pinus sylvestris* | Finnish forests | | | roots | 880–1970 | 12 |
| | | | | needles | 270–320 | 12 |
| *Pinus sylvestris* | Finnish forests | | | needles | 190–300 | 11 |
| *Populus tremula* | Finnish forests | | | leaves | 39 | 11 |
| *Acer saccharum* | NS | | 0 | stems | 200 | 13 |
| | | | 27 | stems | 300 | 13 |
| | | | 54 | stems | 400 | 13 |
| | | | 0 | roots | 800 | 13 |
| | | | 27 | roots | 4000 | 13 |
| | | | 54 | roots | 10,000 | 13 |
| *Betula pubescens* | Finnish forests | | | leaves | 15–130 | 13 |
| *Lotus pedunculatus* | soil solution | 4.7 | 0 | shoots | 98 | 14 |
| | | | 270 | shoots | 113 | 14 |
| | | | 0 | roots | 370 | 14 |
| | | | 270 | roots | 755 | 14 |
| *Lotus corniculatus* | soil solution | 4.7 | 0 | shoots | 108 | 14 |
| | | | 270 | shoots | 217 | 14 |
| | | | 0 | roots | 355 | 14 |
| | | | 270 | roots | 792 | 14 |

| | | | | | | |
|---|---|---|---|---|---|---|
| *Polytrichium juniperinum* | forests | | | foliage | 5500 | 5 |
| *Pleurozium schreberi* | Finnish forests | | | shoots | 130–813 | 11 |
| *Lycopodium* sp. | forests | | | foliage | 26,500 | 5 |
| *Equisetum arvense* | forests | | | foliage | 10,800 | 5 |
| *Hydrangea macrophylla* | forests | | | foliage | 13,000 | 5 |
| *Vochysia thrysoides* | forests | | | foliage | 14,100 | 5 |
| *Qualea multiflora* | forests | | | foliage | 11,500 | 5 |
| *Palicourea rigida* | forests | | | foliage | 9900 | 5 |
| *Vaccinium myrtillus* | Finnish forests | | 43–46[d] | leaves | 86–168 | 11 |
| *Vaccinium vitis-ideaea* | Finnish forests | | 43–46[d] | leaves | 89–320 | 11 |
| Mushrooms | Finnish forests | | 43–46 | entire | 1–182 | 11 |
| | | B. Cultivated Crops | | | | |
| Bragg soybean | MS | 3.9–4.9 | 66–519[b] | shoots | 107–140 | 15 |
| Essex soybean | MS | 3.9–4.9 | 66–519[b] | shoots | 83–161 | 15 |
| *Vicia villosa* | MS | 3.9–4.9 | 66–519[b] | shoots | 227–319 | 15 |
| *Trifolium incarnatum* | MS | 3.9–4.9 | 66–519[b] | shoots | 259–350 | 15 |
| *Camelia sinensis* | | | | leaves | 30,700 | 16 |

Habitat: FW, freshwater; M, marine; SC, sand culture; NS, nutrient solution; MS, mine spoils. Al in habitat: concentration in [a]water, [b]soil, [c]pore water, [d]humus. Tissue: entire = whole body. n.d., no data, b.d., below detection limits.

References: (1) Havas 1986a,b; (2) Miles, unpublished data; (3) Sprenger and McIntosh 1989; (4) Miller et al. 1983; (5) Freedman and Hutchinson 1986; (6) Lehtonen 1989; (7) Mueller et al. 1987; (8) Folkeson et al. 1990; (9) Schier 1985; (10) Hutchinson et al. 1986; (11) Nuorteva 1990; (12) Vilkka et al. 1990; (13) Thornton et al. 1986; (14) Blamey et al. 1990a; (15) Taylor et al. 1992; (16) Matsumoto et al. 1976.

in *Vaccinium myrtilloides* leaves were 2.3–3.1 times greater in autumn than in spring. As part of the same large study, Vilkka et al. (1990) reported greater levels in roots and needles of *Pinus sylvestris* during June than August in rural areas. However, seasonal differences were less pronounced in urban areas, where the source of contamination was probably more consistent.

Aluminum levels in plants may decline rapidly with distance from a point source, but the total amount and rate of decline may depend on the prevailing direction of wind. Mueller et al. (1987) measured metal levels in lichens around a smelter in Louisiana. They found that levels decreased exponentially with distance from the source, with a half-rate of approximately 12.3 km. Airborne dust from gravel roadways can add substantially to Al in *Sphagnum* spp., but levels drop logarithmically with distance from roads (Santlemann and Gorham 1988). The metal in needles of *Pinus sylvestris* was higher in urban environments where airborne contamination was presumably higher than in rural settings, but its levels in roots did not differ in the two sampling sites (Vilkka et al. 1990).

Terrestrial species of plants seem to store greater concentrations of Al in roots than in other tissues. Roots of red, white, and black spruce grown in sand culture had 5.0–10.9 times more than their shoots (Hutchinson et al. 1986). Roots of *Pinus sylvestris* had 3.2–6.1 times as much as needles (Vilkka et al. 1990), and *Lotus* spp. roots accumulated 1.1–6.7 times as much as shoots (Blamey et al. 1990b). Hyperaccumulators among noncultivated terrestrial plants include *Picea rubens, Picea glauca, Picea mariana,* and *Pinus banksiana* (Hutchinson et al. 1986) and several species of shrubs (Freedman and Hutchinson 1986; Haridasan 1982) (Table 4). Among cultivated plants, tea excels as a hyperaccumulator with levels exceeding 30,000 $mg \cdot kg^{-1}$ in leaves (Matsumoto et al. 1976).

## E. Hazard Assessment

Most studies have shown that aquatic plants are more tolerant of Al than aquatic animals. Although certain sensitive algae species may be replaced by more tolerant forms, overall biomass and primary productivity are not consistently different in ponds or streams with elevated Al or acidity. Relationships between individual algae species and higher trophic organisms are not completely understood, but the considerable variation often seen among ponds suggests that overall biomass and productivity of green algae is more important than algal species composition in nutrient and mineral exchange.

Terrestrial plants are potentially at much higher risk from elevated Al than aquatic plants, especially where soils are acidic. Soil levels of the metal are measured in thousands of parts per million rather than hundredths as in aquatic systems, and reduced soil pH can alter the species of Al to more toxic forms. Although most attention has focused on crop species and secondarily on trees, the distribution and vigor of other native upland species also may be limited by soil Al. In particular, species that have evolved with high-Al soils and high soil pH may be at considerable risk due to acid deposition in parts of the eastern U.S. and Canada; this would include red spruce, white spruce, balsam fir, *Populus* spp., and peach. Elevated Al may be a factor in the die-backs observed in red spruce and balsam fir in eastern North America. There is less information on sugar maple, but this species appears to be more tolerant and its die-backs may result from other factors.

## IV. Effects on Invertebrates

Most studies of Al toxicity on invertebrates concern freshwater species, especially crustaceans and larval or nymphal stages of insects. We were unable to find any references to toxicity in terrestrial invertebrates and only one on a strictly estuarine species. Although some studies have examined effects of Al at pHs around neutral, most have focused on the relationships between acid deposition, Al, and aquatic organisms.

### A. Toxicity

Aluminum tends to be nontoxic to invertebrates at aqueous levels commonly found in circumneutral water. Above pH 5.5, lethal concentrations may exceed 23 $mg \cdot L^{-1}$ (Table 5). In contrast, ambient levels of Al in water at circumneutral pH are usually less than 1 $mg \cdot L^{-1}$ and more typically around 500 $\mu g \cdot L^{-1}$ (Wren and Stephenson 1991).

In many studies, adding Al to acidified water produced few differences other than acidity alone. Exposure to 400–500 $\mu g \cdot L^{-1}$ at a pH of 4.0 had no effect beyond that of $H^+$ on clams (*Pisidium* sp.), amphipods (*Hyalella* sp.), isopods (*Asellus* sp.), snails (*Amnicola, Physella* sp.), or insect larvae (*Enallagma* sp., *Lepidostoma* sp., *Pycnopsyche* sp.) (Burton and Allan 1986; Mackie 1989). Similarly, exposure to pH 4.28 and 350 $\mu g \cdot L^{-1}$ Al did not increase mortality of larval *Chironomus* sp., *Hydropsyche* sp., or *Dinocras* sp. (Ormerod et al. 1987b). Levels as high as 1000 $\mu g \cdot L^{-1}$ did not kill *Chaoborus* sp., *Chironomus* sp., or *Holopedium* sp. at pH 4.0, although mortality was greater for *Holopedium* sp. at pH 4.0 than at circumneutrality (Havas and Likens 1985).

Slight (~5%) increases in mortality due to Al have been observed for the larvae of *Ecdyonorus* sp. and *Baetis* sp. at 350 $\mu g \cdot L^{-1}$ and pH 4.3 (Ormerod et al. 1987b) and in the amphipod *Gammarus* sp. exposed to a pH of 4.9 and 270 $\mu g \cdot L^{-1}$ Al (McCahon and Pascoe 1989). At a pH of 4.7 and 890 $\mu g \cdot L^{-1}$, larval *Rhyacophila* sp. and *Dinocras* sp. showed 20% mortality, but none perished at the same pH without Al (McCahon et al. 1989). Season may have some influence on toxicity in that the addition of 500 $\mu g \cdot L^{-1}$ Al to water of pH 4.0 had no additive effect on mortality in the stoneflies *Nemoura* sp. or *Asellus* sp. during summer but did increase mortality in winter (Burton and Allan 1986).

Table 5. Acute, chronic, and sublethal effects of aluminum on aquatic invertebrates.

| Taxon | Method | Chemical | Hardness ($mg \cdot L^{-1}$ $CaCO_3$) | pH | $EC_{50}$ or $LC_{50}$ | Response | Reference |
|---|---|---|---|---|---|---|---|
| | | | A. Acute Toxicity | | | | |
| Planaria | | | | | | | |
| *Dugesia tigrina* | A,S | $AlCl_3$ | 47.4 | 7.5 | 23,000 | $LC_{50}$[a] | 1 |
| Gastropoda | | | | | | | |
| *Physa* sp. | A,S | $AlCl_3$ | 47.4 | 7.5 | 55,000 | $LC_{50}$ | 1 |
| | | | | 6.6 | >23,400 | $LC_{50}$ | 1 |
| | | | | 7.5 | 30,600 | $LC_{50}$ | 1 |
| | | | | 8.2 | 24,700 | $LC_{50}$ | 1 |
| *Amnicola limosa* | A,S | Al wire | 15.3 | 3.5 | >1000 | $LC_{50}$[b] | 2 |
| Pelecypoda | | | | | | | |
| *Pisidium casertanum* | A,S | Al wire | 15.3 | 3.5 | >1000 | $LC_{50}$[b] | 2 |
| | | | | 4.0 | >400 | $LC_{50}$[b] | 2 |
| | | | | 4.5 | >400 | $LC_{50}$[b] | 2 |
| *Pisidium compressum* | A,S | Al wire | 15.3 | 3.5 | >1000 | $LC_{50}$[b] | 2 |
| | | | | 4.0 | >400 | $LC_{50}$[b] | 2 |
| | | | | 4.5 | >400 | $LC_{50}$[b] | 2 |
| Cladocera | | | | | | | |
| *Ceriodaphnia dubia* | A,S | $AlCl_3$ | 50.0 | 7.4 | 1900 | $LC_{50}$[c] | 3 |
| *Ceriodaphnia* sp. | A,S | $AlCl_3$ | 47.4 | 7.7 | 3690 | $LC_{50}$[d] | 1 |
| *Daphnia magna* | A,S | $AlCl_3$ | 45.3 | 6.5–7.5 | 3900 | $LC_{50}$ | 4 |
| | A,S | $AlCl_3$ | 45.4 | 7.6 | >25,000 | $LC_{50}$ | 1 |
| | A,S | $Al_2(SO_4)_3$ | 220 | 7.0 | 38,200 | $LC_{50}$ | 1 |
| | A,S | $AlCl_3$ | 2.5 Ca | 6.5 | 3200 | mortality 40% greater compared to pH alone | 5 |

| | | | | | | | |
|---|---|---|---|---|---|---|---|
| *Daphnia middendorffiana* | A,S | $AlCl_3$ | NA | 4.5 | 20,000 | 100% mortality after 20 hr | 6 |
| *Daphnia catawba* | A,S | $AlCl_3$ | NA | 6.5 | 1020 | significant increase in mortality | 7 |
| Amphipoda | | | | | | | |
| *Gammarus pseudolimnaeus* | A,S | $AlCl_3$ | 47.4 | 7.5 | 22,000 | $LC_{50}$ | 1 |
| *Hyalella azteca* | A,S | Al wire | 15.3 | 5.0 | >1000 | $LC_{50}$[b] | 2 |
| Insecta | | | | | | | |
| Stonefly: *Acroneuria* sp. | A,S | $AlCl_3$ | 47.4 | 7.5 | >22,000 | $LC_{50}$ | 1 |
| Midge: *Tanytarsus dissimilis* | A,S | $Al_2(SO_4)_3$ | 17.4 | 6.8–7.7 | >79,900 | $LC_{50}$ | 8 |
| Odonate: *Enallagma* sp. | A,S | $AlCl_3$ | 15.3 | 3.5 | >1000 | $LC_{50}$ | 2 |
| | | | | 4.0 | >400 | $LC_{50}$ | 2 |
| | | | | 4.5 | >400 | $LC_{50}$ | 2 |
| Ephemeroptera: | | | | | | | |
| *Baetis* sp. | A,I | $Al_2(SO_4)_3$ | 20.7 | 5.0 | 347 | 5% increase in mortality | 9 |
| | A,I | $Al_2(SO_4)_3$ | 4.9 Ca | 4.9 | 270 | 78% decline in density | 10 |
| *Ecdyonorus* sp. | A,I | $Al_2(SO_4)_3$ | 20.7 | 5.0 | 347 | 5% increase in mortality | 9 |
| *Rhyacophila* sp. | A,I | $Al_2(SO_4)_3$ | 1.8 Ca | 4.7 | 890 | 20% mortality | 11 |
| *Dinocras* sp. | A,I | $Al_2(SO_4)_3$ | 1.8 Ca | 4.7 | 890 | 20% mortality | 11 |
| Benthic invertebrates | A,I | $Al_2(SO_4)_3$ | 4.9 Ca | 4.9 | 270 | 60% decline in density | 10 |

*(continued)*

Table 5. (*Continued*)

| Taxon | Method | Chemical | Hardness ($mg \cdot L^{-1}$ $CaCO_3$) | pH | $EC_{50}$ or $LC_{50}$ | Response | Reference |
|---|---|---|---|---|---|---|---|
| | | | B. Chronic Toxicity | | | | |
| Cladocera | | | | | | | |
| *Ceriodaphnia dubia* | C | $AlCl_3$ | 50 | 7.1 | 1908 | $LC_{50}$ | 3 |
| *Daphnia magna* | C | $Al_2(SO_4)_3$ | 220 | 8.3 | 742 | $LC_{50}$ | 1 |
| *Daphnia magna* | C | $AlCl_3$ | 45.3 | 7.7 | 1400 | $LC_{50}$ | 4 |
| *Daphnia galeata* | C,I | $Al_2(SO_4)_3$ | NA | 4.5 | 240 | 100% mortality 13 d earlier than pH alone | 12 |
| | | | C. Sublethal Effects | | | | |
| Cladocera | | | | | | | |
| *Daphnia magna* | C | $AlCl_3$ | 45.3 | 7.7 | 680 | 50% reproductive impairment | 4 |
| Decapoda (crayfish) | | | | | | | |
| *Astacus astacus* | C,F | $Al_2(SO_4)_3$ | 40.1 | 4.0 | 674 | in hemolymph: reduced $PCO_2$ and K; increased Ca and $PO_2$ | 13 |
| *Oronectes virilis* | C,S | $AlCl_3$ | 11.0 | 7.0 | 200 | no effect | 14 |
| | | | | 5.5 | 200 | decreased Ca uptake by 47% | 14 |

| | | | | | | | |
|---|---|---|---|---|---|---|---|
| Insecta | | | | | | | |
| Mayfly: *Heptagenia* sp.; | A,S | $Al_2(SO_4)_3$ | 2.8 mg Ca | 4.0 | 2000 | increased respiration rate, 60% | 15 |
| *Ephemera* sp. | | | | 4.8 | 2000 | increased respiration rate, 93% | 15 |
| Odonate: *Somatochlora* sp. | A,F | $AlCl_3$ | NA | 4.2 | 20,000 | decreased respiration rate, 23–38%; and ammonia excretion 10–42% | 11 |

Methods: A, acute; C, chronic; S, static; F, flow through; T, translocation.

Response: a, 48 hr; b, 72 hr; c, <16 hr; d, <24 hr.

References: (1) Gostomski 1990; (2) Mackie 1989; (3) McCauley et al. 1986; (4) Biesinger and Christensen 1972; (5) Havas 1985; (6) Havas and Hutchinson 1982; (7) Havas and Likens 1985; (8) Lamb and Bailey 1981; (9) Ormerod et al. 1987b; (10) Weatherley et al. 1988; (11) Correa et al. 1985; (12) Havens and Heath 1989; (13) Jensen and Malte 1990; (14) Malley and Chang 1985; (15) Herrmann and Andersson 1986.

Cladocerans appear to be sensitive to Al under acid conditions (Havas and Hutchinson 1982); these organisms also tend to be among the most sensitive to low pH. In some cladocerans, Al may temporarily ameliorate $H^+$ toxicity by interfering with ion transport and reducing the loss of Na and Cl. These organisms experienced severe mortality at pH below 5.0, and elevated Al appeared to prolong life. However, all of the organisms tested at low pHs were dead after 24 hr, regardless of Al concentration. Temporary respite from $H^+$ toxicity has also been observed in the amphipod *Gammarus* sp. (McCahon et al. 1989; Ormerod et al. 1987b). At circumneutral pH, 1020 $\mu g \cdot L^{-1}$ Al significantly increased mortality in cladocerans (Havas and Hutchinson 1982).

### B. Sublethal Effects

Below pH 5.5, Al may affect ion regulation and Ca uptake in crustaceans and insects. Reduced pH by itself negatively affects Na, Cl, and K balance in crustaceans, although insect larvae may be more tolerant (Havas and Hutchinson 1983). The greatest loss of Na and Cl in experimentally treated *Daphnia magna* occurred when they were exposed to low pH ($\leq 5.0$) and elevated Al ($\geq 320$ $\mu g \cdot L^{-1}$) in soft water (Havas 1985). At pH 7.0, 200 $\mu g \cdot L^{-1}$ had no effect on Ca uptake in the crayfish *Orconectes virilis* (Malley and Chang 1985). When the pH was reduced to 5.5, Ca uptake was lowered to 30% of its circumneutral level, and the addition of 200 $\mu g \cdot L^{-1}$ at pH 5.5 reduced Ca uptake by another 10%. In comparison to water at pH 4.0 with no added Al, 674 $\mu g \cdot L^{-1}$ Al increased hemolymph Ca and $PO_4$ and Al content in gills but reduced hemolymph K in the crayfish *Astacus astacus* (Jensen and Malte 1990).

Changes in respiration rates and oxygen transport have been observed in some invertebrates subjected to elevated Al and acidity. Nymphs of the dragonfly *Somatochlora cingulata* experienced a 23%–38% drop in oxygen consumption and a 10%–42% drop in ammonia ($NH_3$) excretion when exposed to pH 4.2 and 20 $mg \cdot L^{-1}$ Al, compared to nymphs at the same pH but without added Al (Correa et al. 1985). Oxygen consumption declined by 30%–50% in the acid plus Al treatment compared to controls. The effects of acid and Al increased as body size of insects decreased. Respiration rates in three species of mayflies increased with 2000 $\mu g \cdot L^{-1}$ Al at pH 4.0–4.8 compared to control animals (Herrmann and Andersson 1986), but the study did not adequately test the effects of Al independently of acidity.

### C. Community Effects

A few studies have reported significant increases in insect drift rates from acidified sections of streams. In some cases, additions of the metal have increased drift compared to acid alone. Ormerod et al. (1987b) showed that drift rates of the mayfly *Baetis rhodani* increased 8.4-fold in a section of stream treated with 347 $\mu g \cdot L^{-1}$ Al and acid (pH 4.3), whereas drift in an acidified section increased only 1.4-fold; both values are in comparison to a control (pH 7.0) upstream portion. There was also a significant reduction in the density of *B. rhodani* in the benthos of the section treated with acid and Al. Other invertebrate larvae (*Dixa puberula, Protonemura meyeri, Ephemeralla ignita*, and *Dicranota* sp.; Simulidae) and total invertebrates also drifted at a greater rate in the acid with Al than in the acid-only section. Drift of *Baetis rhodani* increased in acidified portions of a stream experiencing natural inputs of Al and acid, but no difference was detectable between Al (pH 4.9, 270 $\mu g \cdot L^{-1}$) and acid-only portions of the stream (Weatherley et al. 1988).

In other *in situ* experiments, mortality of *B. rhodani* and *Gammarus pulex* showed a slight but statistically significant increase with the addition of 270–890 $\mu g \cdot L^{-1}$ Al at pH 4.7–4.9 (McCahon et al. 1989). No differences in invertebrate biomass or density due to Al were detectable in experimental stream channels treated with 190 $\mu g \cdot L^{-1}$ for 73 d compared to channels treated only with acid (pH 4.0). Comparative studies (McCahon et al. 1989; Ormerod et al. 1987b) clearly demonstrate that fish are more sensitive than are aquatic invertebrates. As a result, elevated Al may enhance invertebrate numbers by reducing the number of piscine predators.

### D. Bioaccumulation

In addition to being directly affected by acidification and elevated Al, aquatic invertebrates may accumulate Al on or within their bodies. Consequently, they may concentrate the metal to levels that are toxic to predators. Body burdens under typical ambient levels of Al frequently exceed 1000 $mg \cdot kg^{-1}$ and may exceed 16,000 $mg \cdot kg^{-1}$ in experimental situations (Table 6). Among free-living freshwater insects, Al has been recorded as high as 4900 $mg \cdot kg^{-1}$ (Sadler and Lynam 1985).

*Daphnia magna* accumulated more than 2000 $mg \cdot kg^{-1}$ of Al when immersed in water with 1020 $\mu g \cdot L^{-1}$ for 24 hr (Havas 1985); most of this Al was adsorbed onto gills and outer shell surfaces. Bioconcentration factors ranged from 0 at pH 4.5, where most died within a short period, to 10,000-fold at pH 6.5. When the mayfly *Heptagenia sulphurea* was exposed to water with 2000 $\mu g \cdot L^{-1}$ Al for 4 wk, its bioconcentration factor was 1170-fold with a final body concentration of 2340 $mg \cdot kg^{-1}$ (Frick and Herrmann 1990). Seventy percent of this burden was lost with exuviae, indicating that the metal was largely adsorbed rather than assimilated. Similarly, McCahon et al. (1989) determined that exuvia of insect larvae in waters with elevated Al contained 91 times more Al than that of recently molted larvae. Frick and Herrmann (1990) speculated that terrestrial predators had little risk of contamination by aquatic insects because most species except Plecoptera molt before emerging as adults. However, species that feed on larval insects may be exposed to high dietary levels of the metal.

Very high levels of Al were found in the tissues of benthic invertebrates

Table 6. Aluminum concentrations ($mg \cdot kg^{-1}$) in invertebrates under natural and experimental conditions.

| Taxon | Tissue | Level ($mg \cdot kg^{-1}$) | Comments | Reference |
|---|---|---|---|---|
| Composite benthic sample | whole body | 2396 | north Chesapeake Bay, MD; high 8900 $mg \cdot kg^{-1}$ | 1 |
| **Mollusca** | | | | |
| Gastropoda | whole body | 27–398 | natural ponds in MA and MD; pH 5.7–5.8 | 2 |
| Baltic clam, *Macoma* sp. | whole body | 1994 | north Chesapeake Bay, MD | 1 |
| | | 3010 | south Chesapeake Bay, MD | 1 |
| | | 12,475 | San Francisco Bay, CA | 1 |
| Ribbed mussel, *Geukensia* sp. | whole body | 450 | South Chesapeake Bay, MD | 1 |
| | | 1435 | San Francisco Bay, CA | 1 |
| **Crustacea** | | | | |
| *Daphnia middendorffiana* | whole body | 5100 | pH 4.0, low-Al water | 3 |
| *Daphnia magna* | whole body | 360 | experimental, pH 6.5, Al 200 $\mu g \cdot L^{-1}$ | 4 |
| | whole body | 14,974 | pH 6.5, Al 1020 $\mu g \cdot L^{-1}$ | 4 |
| | whole body | 180 | pH 5.0, Al 200 $\mu g \cdot L^{-1}$ | 4 |
| | whole body | 3366 | pH 5.0, Al 1020 $\mu g \cdot L^{-1}$ | 4 |
| | whole body | 86 | pH 4.5, Al 200$\mu g \cdot L^{-1}$ | 4 |
| | whole body | 1630 | pH 4.5, Al 1020 $\mu g \cdot L^{-1}$ | 4 |
| *Gammarus pulex* | whole body | 470 | natural ponds in Wales | 5 |
| | whole body | 300–900 | natural streams in Wales, pH 7.2–7.4, Al 100–200 $\mu g \cdot L^{-1}$ | 6 |
| | whole body | 120 | experimentally limed stream, pH 6.5, Al 130 $\mu g \cdot L^{-1}$ | 7 |
| *Isoperla grammatica* | whole body | 50 | experimentally limed stream as above | 7 |
| *Procambarus clarkii* | abdominal muscle | 1.22 | crayfish farm | 8 |
| | hepatopancreas | 1.42 | crayfish farm | 8 |
| | alimentary canal | 26.9 | crayfish farm | 8 |
| | exoskeleton | 4.28 | crayfish farm | 8 |

| | | | | |
|---|---|---|---|---|
| | blood | 37.9 | crayfish farm | 8 |
| | abdominal muscle | 1.75–6.39 | roadside ditches | 8 |
| | hepatopancreas | 3.10–22.74 | roadside ditches | 8 |
| | alimentary canal | 309–981 | roadside ditches | 8 |
| | exoskeleton | 10.8–77.4 | roadside ditches | 8 |
| | blood | 35.9–140 | roadside ditches | 8 |
| *Astacus astacus* | gills | 1484 | experimentally exposed to 674 $\mu g \cdot L^{-1}$, pH 4.0 for 21 d | 9 |
| | muscle | 485 | as above | |
| | gills | 1079 | exposed to pH 4.0 without added Al | |
| | muscle | 485 | as above | |
| *Orconectes virilis* | carapace | 65.2 | freshly collected | 10 |
| | hepatopancreas | 19.6 | freshly collected | 10 |
| | green gland | 84.4 | freshly collected | 10 |
| | gills | 32.9 | freshly collected | 10 |
| | gut | 774 | freshly collected | 10 |
| | ovary | 50.4 | freshly collected | 10 |
| | muscle | 6.1 | freshly collected | 10 |
| | other | 71.7 | freshly collected | 10 |
| | carapace | 46.9 | fasted in 500 $\mu g \cdot L^{-1}$ Al solution, 14 d | 10 |
| | hepatopancreas | 9.1 | as above | 10 |
| | green gland | 105.6 | as above | 10 |
| | gills | 38.3 | as above | 10 |
| | gut | 89.0 | as above | 10 |
| | ovary | 13.8 | as above | 10 |
| | muscle | 4.4 | as above | 10 |
| | other | 24.0 | as above | 10 |
| Shore crab, *Sesarma* sp. | whole body | 1847 | south Chesapeake Bay, MD | 1 |
| Shore crab, *Pachygrapsus* sp. | whole body | 1965 | San Francisco Bay, CA | 1 |

(*continued*)

Table 6. (*Continued*)

| Taxon | Tissue | Level (mg·kg$^{-1}$) | Comments | Reference |
|---|---|---|---|---|
| | | **Insecta** | | |
| Composite samples | whole body | 26–700 | samples from MD, pH 5.7–5.8 | 2 |
| | | 4–240 | samples from MA, pH 5.7–5.8 | 2 |
| Hemiptera | | | | |
| *Gerris lacustris* | whole body | 10–16 | lakes in Finland | 11 |
| Diptera | whole body | 1900 | natural streams in Wales, pH 4.5–8.2, Al 23–500 $\mu$g·L$^{-1}$ | 5 |
| Chironomidae | whole body | 100–3300 | natural streams in Wales, pH 5.0–8.1, Al b.d.[a]-1600 $\mu$g·L$^{-1}$ | 6 |
| Chironomidae | whole body | 650–2500 | lakes in Finland | 11 |
| *Chironomous* sp. | whole body | 5000 | natural sediments with 500 mg·kg$^{-1}$ Al | 12 |
| Odonata | | | | |
| *Aeschna* sp. | imago | 100–890 | lakes in Finland | 11 |
| Coenagriidae | imago | 20–150 | lakes in Finland | 11 |
| *Lestes* sp. | imago | 77–170 | lakes in Finland | 11 |
| *Coenagrion* sp. | imago | 53–76 | lakes in Finland | 11 |
| *Cordulia metallica* | imago | 30–37 | lakes in Finland | 11 |
| Plecoptera | whole body | 1270 | Welsh streams, pH 4.5–8.2, Al 23–500 $\mu$g·L$^{-1}$ | 5 |
| | whole body | 200–4900 | Welsh streams, pH 5.0–8.1, Al b.d.–1600 $\mu$g·L$^{-1}$ | 6 |
| Ephemeroptera | whole body | 2130 | Welsh streams, pH 4.5–8.2, Al 23–500 $\mu$g·L$^{-1}$ | 5 |
| | whole body | 100–4900 | Welsh streams, pH 5.0–8.1, Al b.d.–1600 $\mu$g·L$^{-1}$ | 6 |
| Trichoptera | whole body | 1590 | Welsh streams, pH 4.5–8.2, Al 23–500 $\mu$g·L$^{-1}$ | 5 |
| | whole body | 100–2300 | Welsh streams, pH 5.0–8.1, Al b.d.–1600 $\mu$g·L$^{-1}$ | 6 |

| | | | | |
|---|---|---|---|---|
| *Heptagenia sulphurea* | whole body | 50 | Experimental, Al 200 $\mu g \cdot L^{-1}$, pH 4.5 | 13 |
| | | 2340 | Al 2.0 $mg \cdot L^{-1}$, pH 4.5 | 13 |
| | imago/exuvium | 70/670 | Al 200 $\mu g \cdot L^{-1}$, pH 4.5 | 13 |
| | imago/exuvium | 120/10,970 | Al 2.0 $mg \cdot L^{-1}$ pH 4.5 | 13 |
| | imago/exuvium | 0/0 | Al 0.0 $\mu g \cdot L^{-1}$, pH 7.0 | 13 |
| *Limnephilus* sp. | imago | 24–28 | lakes in Finland | 11 |
| *Agryphia* sp. | imago | 42–48 | lakes in Finland | 11 |
| Lepidoptera | | | | |
| *Sphnix pinastri* | larva | 80 | Al 200 $mg \cdot kg^{-1}$ in food plants | 11 |
| *Dendrolimus pini* | larva | 70 | as above | 11 |
| *Panolis flammea* | larva | 6 | as above | 11 |
| *Thera obeliscata* | larva | 25 | as above | 11 |
| *Laothoe populi* | larva | 44 | Al 39 $mg \cdot kg^{-1}$ in food plants | 11 |
| *Achyla flavicornis* | larva | 28 | Al 26 $mg \cdot kg^{-1}$ in food plants | 11 |
| *Epirrita autumnata* | larva | 46 | as above | 11 |
| *Amphopoea oculata* | N/A | 18 | forests in Finland | 11 |
| *Eulithis populata* | N/A | 0–15 | forests in Finland | 11 |
| Coleoptera | | | | |
| *Hylobius abietis* | adults | 32–67 | forests in Finland | 11 |
| Ipidae: | | | | |
| *Tomicus piniperda* | adults | 48–140 | forests in Finland | 11 |
| *Hylurgops palliatus* | adults | 68–94 | forests in Finland | 11 |
| *Ips typographus* | adults | 28 | forests in Finland | 11 |
| *Trypodendron lineatum* | adults | 30–56 | forests in Finland | 11 |
| Carabidae: | | | | |
| *Carabus* sp. | adults | 10–39 | forests in Finland | 11 |
| *Cychrus caraboides* | adults | 14–27 | forests in Finland | 11 |

*(continued)*

Table 6. (*Continued*)

| Taxon | Tissue | Level ($mg \cdot kg^{-1}$) | Comments | Reference |
|---|---|---|---|---|
| Silphidae: | | | | |
| *Necrophorus investigator* | adults | 6–31 | forests in Finland | 11 |
| *Deceoptoma thoracica* | adults | 7–12 | forests in Finland | 11 |
| Calliphoridae: | | | | |
| *Lucilia illustris* | adults | 21 | forests in Finland | 11 |
| *Calliphora loewi* | adults | 100 | forests in Finland | 11 |
| *Calliphora vomitoria* | adults | 28 | forests in Finland | 11 |
| Hymenoptera | | | | |
| *Formica* sp. | adults | 40–79 | forests in Finland | 11 |
| Bombidae: | | | | |
| *Bombus lucorum* | adults | 36–41 | forests in Finland | 11 |
| *Bombus pascuorum* | adults | 19–39 | forests in Finland | 11 |
| **Annelida** | | | | |
| *Dendrobaena octaedra* | adult, whole body | 63–81 | Al 2900 $mg \cdot kg^{-1}$ in humus | 11 |
| Enchytraeidae | adult, whole body | 310–2500 | Al 2900 $mg \cdot kg^{-1}$ in humus | 11 |

Referencess: (1) Miles (unpublished data); (2) Albers and Camardese 1993; (3) Havas and Hutchinson 1983; (4) Havas 1985; (5) Ormerod et al. 1988; (6) Sadler and Lynam 1985; (7) McCahon et al. 1989; (8) Madigosky et al. 1991; (9) Jensen and Malte 1990; (10) Malley & Chang 1985; (11) Nuorteva 1990; (12) Krantzberg and Stokes 1989; (13) Frick and Herrmann 1990.

[a]b.d. below detection limits.

inhabiting industrialized sections of San Francisco and Chesapeake bays (Miles, unpublished data) (Table 6). In Chesapeake Bay, mixtures of benthic invertebrates had a mean of approximately 2400 $mg \cdot kg^{-1}$ Al with a high of 8900 $mg \cdot kg^{-1}$. *Macoma* sp. clams had mean levels of 2000–3000 $mg \cdot kg^{-1}$ and highs of 4100–5100 $mg \cdot kg^{-1}$. The ribbed mussel *Geukensia* sp. had comparatively low levels of 450 $mg \cdot kg^{-1}$. In San Francisco Bay, *Macoma* sp. had mean levels of nearly 12,500 $mg \cdot kg^{-1}$ with highs of 17,600 $mg \cdot kg^{-1}$, and the ribbed mussel had three times as much Al ($\bar{x}$ = 1,435 $mg \cdot kg^{-1}$) as in Chesapeake Bay.

Unfortunately, there seems to be little correspondence between Al levels in invertebrates and either water concentrations or pH. Surveys such as those by Sadler and Lynam (1985), Ormerod et al. (1988), and Albers and Camardese (1993) showed little correlation among these parameters. One area of future research interest, however, is that P levels in insects examined by Ormerod et al. (1988) and Sadler and Lynam (1985) were inversely related to Al levels. This may reflect Al–P binding in aquatic insects.

### E. Hazard Assessment

As with other aquatic organisms, the effects of Al on invertebrates are confounded by pH and may vary depending on species and other factors of water chemistry. Thus, it is difficult to discern independent effects of the metal. In many instances, species that appear tolerant to reduced pH are also tolerant to elevated Al. Aluminum-tolerant organisms include most molluscs, many amphipods and isopods, and certain insect larvae, particularly *Chironomus* sp. and *Chaoborus* sp. Interestingly, although molluscs may be Al tolerant, they are among the taxa notably absent from acidic waters because of inadequate Ca. Some of the more sensitive aquatic species include *Daphnia* sp., mayflies, stoneflies, and caddisflies. These are often very important organisms in aquatic food chains, and their absence may negatively affect productivity and complexity of higher trophic levels, including fish and waterbirds. Reduced food availability would be another factor hampering species of fish that are already stressed by low pH and elevated Al.

Levels in aquatic invertebrates are generally below levels of concern for higher trophic levels. Scheuhammer (1991) suggested that dietary Al should be at least 1000 $mg \cdot kg^{-1}$ dry weight of food or 50% of dietary P to be toxic in birds or mammals: many reported concentrations in aquatic invertebrates are below that. However, P levels in aquatic insects are often low (e.g., Sadler and Lynam 1985) and Al : P ratios may be above 0.5. We suggest that fish and birds that live in acidic habitats and have diets containing large proportions of benthic invertebrates could be at some risk from Al toxicity.

## V. Effects on Fish

Through combined efforts in North America and Europe, about 150 papers addressing impacts on 35 fish species were published between 1980 and

1992, the primary period covered in this review. Although other papers have been published subsequently, none drastically alter what is known about Al toxicity to fish. Over 70% of the effort was directed toward four salmonid species: rainbow trout (*Oncorhynchus mykiss*), Atlantic salmon (*Salmo salar*), brown trout (*Salmo trutta*), and brook trout (*Salvelinus fontinalis*). These species inhabit areas where watershed soils have low buffering capacity. More than 40 responses to reduced pH and elevated Al levels have been documented; only a few are presented here (see Tables 7–13). Physiological effects of Al have been more extensively studied than acute toxicity, probably because the toxic effects of the metal generally are additive to those of reduced pH.

The tables and text for this section are organized in phylogenetic order by species (American Fisheries Society 1980), and tabulated data are by age group for each species. Standardized terminology is used for anadromous salmonids (Allen and Ritter 1977). For all other species, the term "alevin" refers to the stage between hatching and independence from the yolk sac as the primary source of nutrition and the term "juvenile" refers to the stage between alevin and the end of the first year. Year classes of adults are not given unless the responses of different aged adults were presented.

The species included here are those whose ranges include acid-vulnerable areas and are important sport or commercial species. The fathead minnow (*Pimephales promelas*) also is included because it has been used extensively in laboratory testing and is therefore a good reference animal.

## A. Toxicity to Fish

Most fish species are equally sensitive to the three forms of aqueous Al (Driscoll et al. 1980; Rosseland et al. 1992; Skogheim and Rosseland 1986). However, Palmer et al. (1988) suggested that bluegills (*Lepomis macrochirus*) may be more sensitive to the Al–OH complexes than to $Al^{3+}$.

For many species of fish, toxic effects appear to follow a bimodal model based on declining pH that shifts from asphyxiation to impaired ion regulation (Neville and Campbell 1988; Playle et al. 1989). The relative importance of asphyxiation or ion regulation is dependent on pH, Al level, and the magnitude of the pH change of water irrigating the gills (Playle and Wood 1991). Asphyxiation is manifested by increased gill ventilation rates and volumes, blood $CO_2$, and blood lactate and by reduced blood $O_2$ tension. It often is accompanied by increased gill mucus production, coughing rates, and gill tissue damage (Neville 1985; Neville and Campbell 1988; Playle and Wood 1991). Asphyxiation appears to be the predominant toxicity mechanism in the pH range of 6.5–5.5 (Neville and Campbell 1988).

Impaired ion regulation is marked by losses of electrolytes from the blood and body fluids (Neville and Campbell 1988) and increased $H^+$ and carbon dioxide ($CO_2$) concentrations in the blood (Neville 1985). It appears to predominate in the pH range of 5.5–4.5 (Neville and Campbell 1988).

Several studies conducted during the early 1980s showed that elevated Al and reduced pH could cause asphyxiation and disrupt electrolyte regulation in fish. These studies usually covered only a portion of the pH range that fish may encounter in the wild. Consequently, it was difficult to determine if the two toxic mechanisms are important at different pHs. One of the first studies, designed to determine the pH ranges at which the mechanisms operate, measured changes in ventilation rates, percent $O_2$ saturation of blood, plasma concentrations of $Na^+$ and $Cl^-$, and Al levels in the gills of rainbow trout exposed to pH of 6.5–4.0 and total Al of 76 $\mu g \cdot L^{-1}$ (Neville and Campbell 1988). Ventilation rates reached their maxima and blood $O_2$ their minima at pH 6.1. Ventilation rates rose above normal at pH 4.5 and 4.0 due to increased $CO_2$ in the blood resulting from the increased blood acidity. Blood $O_2$ remained at saturation. Levels of Al in gills rose from 270 $\mu g \cdot g^{-1}$ (pH 6.5) to 500 $\mu g \cdot g^{-1}$ (pH 6.1), then dropped to 94 $\mu g \cdot g^{-1}$ (pH 5.5). The investigators suggested that the asphyxiation response resulted from either of two causes: (1) precipitation of $Al(OH)_3$ on the gill surfaces, causing gill mucification and inflammation which blocked diffusion of $O_2$ and $CO_2$ across gill membranes; or (2) formation of mixed gill-ligand-$Al(OH)_x$ and $Al^{3+}$ complexes that provoked an unidentified biochemical response.

Plasma Na and Cl were unaffected until pH 4.5, at which they dropped progressively below normal. The disruption in electrolyte regulation at lower pH results from Al displacement of $Ca^{2+}$, allowing passive diffusion outward of $Na^+$ and $Cl^-$ and inward movement of $H^+$ through the gills. However, under laboratory conditions and pH ~4.0, Al seems to reduce pH toxicity. At or below this pH, $H^+$ becomes more toxic than Al, which binds to the gills and reduces inward diffusion of $H^+$.

Playle and Wood (1991) exposed juvenile rainbow trout to 138 $\mu g \cdot L^{-1}$ Al at pH 5.1, 4.7, and 4.0 for 6 hr. Gill levels progressively rose with increasing pH, primarily from precipitation of $Al(OH)_2^-$ and $Al(OH)_3$ on gill surfaces. The resulting mucification and inflammation of the gill tissues blocked diffusion of $O_2$ and $CO_2$ through the gills.

Asphyxiation and disrupted electrolyte regulation, however, can occur at lower pH if dissolved Al is near saturation. The highest mortalities in some laboratory studies were reported at moderate acidity and Al levels above saturation (Baker and Schofield 1982). The pH of water expired from juvenile rainbow trout gills is 0.2–0.7 pH units higher than that of inspired water in the pH range of 4.1–5.1 (Playle and Wood 1991). Thus, if dissolved Al is near saturation, the pH change of water passing over the gill membranes can cause Al precipitation and asphyxiation.

Saturated Al solutions at pH ~6.0 can threaten free-living fish populations in locations where an influx of low pH water with high dissolved Al mixes with neutral or slightly acidic waters. Such situations may occur during runoff events when small, acidic, Al-laden streams flow into larger streams or small lakes. Oversaturation of dissolved Al resulting from rapidly increasing pH may have killed adult Atlantic salmon during a spawning run in a Norwegian river (Skogheim et al. 1984). Fish inhabiting the mixing zone of lime applied for acid stream rehabilitation also have suffered osmotic stress and damaged gills (Skogheim et al. 1987) and elevated mortalities (Rosseland et al. 1992; Weatherley et al. 1991). Thus, resource managers should be mindful of the potential dangers to fishery resources of acid stream and lake rehabilitation efforts involving the use of buffering agents.

## B. Species Accounts

**Blueback Herring (*Alosa aestivallis*) and Whitefish (*Coregonus clupeaformis*).** Blueback herring is an anadromous species that spawns during April and May in poorly buffered tributaries of Chesapeake Bay (Klauda and Palmer 1987). Lake whitefish is a freshwater species that is widespread across poorly buffered areas of the eastern Canadian provinces (Holtze and Hutchinson 1989).

Blueback herring alevins (Table 7) were sensitive to Al as low as 55 $\mu g \cdot L^{-1}$ at pH 5.7 and 21 $\mu g \cdot L^{-1}$ at pH 5.0 in 96-hr toxicity tests (Klauda et al. 1987). Exposures to 100 $\mu g \cdot L^{-1}$ for as little as 8 hr appear to reduce survival (Klauda and Palmer 1987).

Lake whitefish alevins (Table 7) are sensitive to similar Al and pH levels as the blueback herring alevins. The survival of lake whitefish alevins exposed to pH levels of 4.5 and 50–200 $\mu g \cdot L^{-1}$ Al was higher than in those exposed to these pH levels with no Al (Holtze and Hutchinson 1989).

**Cutthroat Trout (*Oncorhynchus clarki*) and Rainbow Trout (*O. mykiss*).** Cutthroat and rainbow trout are common in coldwater lakes and streams of acid-vulnerable areas of the Canadian and U.S. intermountain west; however, through introductions, rainbow trout have become ubiquitous in populated areas of North America.

The results of a single study of cutthroat trout (Woodward et al. 1989) suggested that alevins and juveniles may be equally sensitive to Al at lower pHs but that alevins are more sensitive at circumneutral pH (Table 7). Significant alevin mortality occurred during 7-d exposures to pH 6.0 and 5.0 and 50 $\mu g \cdot L^{-1}$ Al, but significant juvenile mortality occurred at pH 5.5 and 5.0 with 50 $\mu g \cdot L^{-1}$. Juvenile growth, however, was affected at pH 6.0 with 50 $\mu g \cdot L^{-1}$. Alevin electrolyte regulation (Table 8) was disrupted at the same Al levels that decreased survival except at pH 5, where Al reduced survival without affecting electrolyte regulation.

Table 7. Percent survival, $LT_{50}$, and sizes of fish exposed to various pH, Ca ($mg \cdot L^{-1}$), and Al ($\mu g \cdot L^{-1}$) levels in laboratory experiments. Blanks for species name, life stage, and experimental conditions within the same study are identical to the value above the blank(s).

| | | Experimental conditions | | | | | | Factor | | | |
|---|---|---|---|---|---|---|---|---|---|---|---|
| Species | Life stage | Temp (° C) | pH | Total Al | Available Al | $Ca^{2+}$ | Exposure time | Survival (%) | $LT_{50}$ (hr) | Length (mm) Weight (mg) | Reference |
| *Alosa aestivalis* | egg[1] | 20 | 5.0 | 0 | 0 | 7.5 | 69 hr[2] | 31 | | | 1 |
| | | | | 21 | 21 | | | 61 | | | |
| | | | | 49 | 43 | | | 51 | | | |
| | | | | 104 | 104 | | | 30 | | | |
| | | | | 230 | 225 | | | 0 | | | |
| | | | 5.7 | 0 | 0 | | | 93 | | | |
| | | | | 55 | 32 | | | 73 | | | |
| | | | | 215 | 110 | | | 19 | | | |
| | | | 6.5 | 0 | 0 | | | 93 | | | |
| | | | | 19 | 14 | | | 86 | | | |
| | | | | 233 | 154 | | | 98 | | | |
| *Alosa aestivalis* | alevin | 21 | 5.6 | 0 | | 7.9 | 8 hr[3] | 85 | | | 2 |
| | | | | 100 | | | | 44 | | | |
| | | | | 0 | | | 12 hr[3] | 75 | | | |
| | | | | 100 | | | | 2 | | | |
| | | | | 0 | | | 24 hr[3] | 51 | | | |
| | | | | 150 | | | | 0 | | | |

(continued)

Table 7. (*Continued*)

| Species | Life stage | Experimental conditions: Temp (° C) | pH | Total Al | Available Al | $Ca^{2+}$ | Exposure time | Factor: Survival (%) | $LT_{50}$ (hr) | Length (mm) Weight (mg) | Reference |
|---|---|---|---|---|---|---|---|---|---|---|---|
| | | 20 | 5.0 | 0 | 0 | 7.5 | 96 hr | 1 | 54.0 | | 1 |
| | | | | 21 | 21 | | | 2 | 37.1 | | |
| | | | | 49 | 43 | | | 0 | 40.5 | | |
| | | | | 104 | 104 | | | 0 | 24.2 | | 1 |
| | | | 5.7 | 0 | 0 | | | 11 | 57.5 | | |
| | | | | 28 | 20 | | | 14 | 67.7 | | |
| | | | | 55 | 32 | | | 7 | 40.6 | | |
| | | | | 108 | 55 | | | 0 | 24.2 | | |
| | | | 6.5 | 0 | 0 | | | 65 | | | |
| | | | | 19 | 14 | | | 75 | | | |
| | | | | 46 | 26 | | | 40 | | | |
| | | | | 126 | 65 | | | 90 | | | |
| *Coregonus clupeaformis* | alevin | 6.5 | 5.4 | 0 | | 4.0 | 12 d | 69.8[b,4] | | | 3 |
| | | | | 50 | | | | 93.3[a] | | | |
| | | | 5.1 | 0 | | | | 85.3[a,b] | | | |
| | | | | 200 | | | | 62.0[c] | | | |
| | | | 4.8 | 0 | | | | 86.0[a,b] | | | |
| | | | | 50 | | | | 92.7[a] | | | |
| | | | | 400 | | | | 71.5[b] | | | |

| | | | | | | | | | |
|---|---|---|---|---|---|---|---|---|---|
| | | | 4.5 | 0 | | | 14.1[f] | | |
| | | | | 50 | | | 70.4[b] | | |
| | | | 4.2 | 0 | | | 0[g] | | |
| | | | | 200 | | | 44.4[e] | | |
| *Oncorhynchus clarki* | egg | 10 | 5.0 | 0 | 1.4 | 7 d | 100 | | 4 |
| | | | | 300 | | | 89 | | |
| | | | 4.5 | 0 | | | 82 | | |
| | | | | 300 | | | 98 | | |
| *Oncorhynchus clarki* | alevin | | 6.0 | 0 | | | 93 | 31 | |
| | | | | 50 | | | 41*[5] | 27* | |
| | | | 5.5 | 0 | | | 99 | 30 | |
| | | | | 100 | | | 31* | 27* | |
| | | | 5.0 | 0 | | | 85 | 30 | |
| | | | | 50 | | | 71* | 29* | |
| | | | | 100 | | | 0* | n/a | |
| *Oncorhynchus clarki* | juvenile | | 6.0 | 0 | | | 100 | 31 | |
| | | | | 50 | | | 100 | 27* | |
| | | | 5.5 | 0 | | | 100 | 31 | |
| | | | | 50 | | | 52* | 26* | |
| | | | 5.0 | 0 | | | 88 | 30 | |
| | | | | 50 | | | 0* | n/a | |

(*continued*)

Table 7. (*Continued*)

| Species | Life stage | Experimental conditions: Temp (° C) | pH | Total Al | Available Al | $Ca^{2+}$ | Exposure time | Factor: Survival (%) | $LT_{50}$ (hr) | Length (mm) Weight (mg) | Reference |
|---|---|---|---|---|---|---|---|---|---|---|---|
| *Oncorhynchus mykiss* | juvenile | 10 | 6.5–5.5 | 0 | | 4.9–2.3 | 6 d | 100 | | | 5 |
| | | | 6.5 | 75 | | 4.9 | | 100 | | | |
| | | | 6.1 | | | 2.8–5.7 | | 32* | | | |
| | | | 5.0 | 0 | | 2.4–3.4 | | 100 | | | |
| | | | | 75 | | 2.1–3.4 | | 100 | | | |
| | | | 4.0 | 0 | | 2.8–3.7 | | $21^{+}$ | | | |
| | | | | 75 | | 2.3–3.7 | | $62^{*+}$ | | | |
| | | 15 | 5.2 | 118 | | 0.6 | 70 hr | | 31.5[6]<br>>70[7] | | 6 |
| *Oncorhynchus mykiss* | adult | | 4.8 | 0 | | 1 | 72 hr | > 72 | | | 7 |
| | | | | 100 | | | | 27.0 | | | |
| | | 13 | 4.3 | 0 | | 38 | 96 hr | 100 | | | 8 |
| | | | | 50 | | | | 83 | | | |
| | | | | 90 | | | 48 hr | 33 | | | |
| | | | | 900 | | | 24 hr | 17 | | | |
| | | | 7.8 | | | | 96 hr | 100 | | | |
| | | | 8.7 | 0 | | | 48 hr | 86 | | | |
| | | | | 50 | | | 96 hr | 100 | | | |
| | | | | 90 | | | 24 hr | 60 | | | |
| | | | | 900 | | | | 83 | | | |
| | | | | | | | 48 hr | 0 | | | |

| | | | | | | | | | | |
|---|---|---|---|---|---|---|---|---|---|---|
| | | 15 | 5.2 | 0 | | 1.8 | 66 hr | 100 | | 9 |
| | | | | | | 16.4 | | 81 | | |
| | | | | 105 | | 1.8 | | 10 | | |
| | | | | | | 16.4 | | 44 | | |
| | | | 4.8 | 0 | | —[8] | | 100 | | |
| | | | | 105 | | 1.8 | | 37 | | |
| | | | | | | 16.4 | | 62 | | |
| *Salmo salar* | egg | | | | | 1–3 | 548 °d | 61.1 | | 10 |
| *Salmo salar* | alevin | 15 | 5.0 | 23 | 5 | | 96 hr | 100 | | 11 |
| | | | | 181 | 135 | | 60 hr | 0 | | |
| | | | | 168 | 111 | | 72 hr | | | |
| | | | | 169 | 41 | | 48 hr | | | |
| | | 18 | 5.1 | 110 | 33.3 | < 1 | | | 150 | 12 |
| | | | | 180 | 59.3 | | | | 93 | |
| | | | | 230 | 77.8 | | | | 62 | |
| | | 8 | 5.5 | < 3 | | | 30 d | 0 | | 13 |
| | | | | 264 | | | 30 d | 85* | | |
| | | | | | | 1–3 | 1409 °d | 6.8 | | 10 |
| *Salmo salar* | parr | 6 | 5.0 | 140 | 44.4 | < 1 | | | 89 | 12 |
| | | | 5.2 | 180 | 59.3 | | | | 60 | |
| | | | 4.9 | 200 | 66.7 | | | | 52 | |
| | | | 4.9 | 250 | 85.2 | | | | 41 | |
| | | | 5.3 | 310 | 107.4 | | | | 27 | |
| | | | 5.2 | 420 | 148.2 | | | | 26 | |
| | | 12 | 5.0 | 290 | 88.9 | | | | 72 | |
| | | | 4.7 | | 96.3 | | | | 43 | |
| | parr[9] | 5.2 | 5.0 | 415 | | | 64 hr | 0 | 27.1 | 14 |
| | parr[10] | | | | | | | 0 | 15.9 | |

*(continued)*

Table 7. (*Continued*)

| Species | Life stage | Experimental conditions: Temp (° C) | pH | Total Al | Available Al | $Ca^{2+}$ | Expo-sure time | Factor: Survival (%) | $LT_{50}$ (hr) | Length (mm) Weight (mg) | Refer-ence |
|---|---|---|---|---|---|---|---|---|---|---|---|
| *Salmo salar* | smolts | | ≤5.6 | 268 | 114 | 1.04 | 48 hr | 0 | | | 15 |
| | | | 6.05 | 267 | 70 | 1.81 | | 30 | | | |
| | | | 6.45 | 268 | 50 | 2.17 | | 100 | | | |
| *Salmo salar* | adult | | 5.0 | 163 | 75 | 1.30 | 18 d | 0 | 109[h] | | 16 |
| | | | | 233 | 137 | | | 0 | 39[h] | | |
| | | | | 277 | 177 | | | 0 | 32[h] | | |
| *Salmo trutta* | egg | | | | | | 521 °d | 0.9 | | | 10 |
| | | | | | | | 540 °d | 39.0 | | | |
| *Salmo trutta* | alevin | 10 | 5.4 | 0 | | 1.0 | 16 d | 100 | | | 17 |
| | | | | 250 | | | | 74 | | | |
| | | | | 500 | | | | 35 | | | |
| | | | 5.1 | 0 | | | | 100 | | | |
| | | | | 250 | | | | 91 | | | |
| | | | | 500 | | | | 46 | | | |
| | | | 4.8 | 0 | | | | 100 | | | |
| | | | | 250 | | | | 95 | | | |
| | | | | 500 | | | | 72 | | | |
| | | | 4.5 | 0 | | | | 100 | | | |
| | | | | 250 | | | | 100 | | | |
| | | | | 500 | | | | 80 | | | |

| | | | | | | | | | | |
|---|---|---|---|---|---|---|---|---|---|---|
| | | | 5.4 | 0 | | 0.50 | | 100 | | |
| | | | | 250 | | | | 24 | | |
| | | | | 500 | | | | 0 | | |
| | | | 5.1 | 0 | | | | 100 | | |
| | | | | 250 | | | | 15 | | |
| | | | | 500 | | | | 0 | | |
| | | | 4.8 | 0 | | | | 100 | | |
| | | | | 250 | | | | 17 | | |
| | | | | 500 | | | | 0 | | |
| | | | 4.5 | 0 | | | | 85 | | |
| | | | | 250 | | | | 0 | | |
| | | 10 | | —[11] | | 0.9 | 70 d[12] | 0 | | 18 |
| | | 0 | | 0 | | 7.3 | | 86 | | |
| | | | | 157 | 149 | 7.0 | | 16 | | |
| | | | 5.5 | 0 | | 0.9 | | 92 | | |
| | | | | 143 | 119 | 0.8 | | 64 | | |
| | | | | 0 | | 7.5 | | 100 | | |
| | | | | 167 | 130 | 7.2 | | 90 | | |
| | | | | | | 1.3 | 1018 °d | 0.3 | | 10 |
| | | | | | | | 1020 °d | 15.0 | | |
| *Salmo trutta* | parr[9] | 5.2 | 5.0 | 410 | | | 64 hr | 30 | 41 | 14 |
| | parr[10] | | | | | | | 60 | | |
| *Salmo trutta* | juvenile | 10 | 4.5 | 324 | | | 12 hr | 92 | | 19 |
| | | | | | | | 30 hr | 55 | | |
| | | | 5 | | | | 60 hr | 26 | | |
| | | | 6.4 | | 99 | 3.07 | 17 d | 60 | 19.5 | 20 |
| | | | 6.3 | | | 5.47 | | 100 | | |
| | | | 5.7 | | | 1.68 | | 100 | | |

(*continued*)

Table 7. (*Continued*)

| | | Experimental conditions | | | | | | Factor | | | |
|---|---|---|---|---|---|---|---|---|---|---|---|
| Species | Life stage | Temp (° C) | pH | Total Al | Available Al | $Ca^{2+}$ | Exposure time | Survival (%) | $LT_{50}$ (hr) | Length (mm) Weight (mg) | Reference |
| | | | 4.9 | | 450 | 2.21 | | 0 | 5.6 | | |
| | | | 4.8 | | 216 | 2.15 | | 20 | 14.0 | | |
| | | | 5.3 | | 333 | 2.00 | | 0 | 8.7 | | |
| | | | | | 126 | 1.24 | 17 d | 60 | 18.0 | | |
| | | | 4.6 | | 216 | 1.36 | | 40 | 13.0 | | |
| | | | 4.4 | | 513 | 1.54 | | 0 | 5.2 | | |
| | adult | 5 | 4.0 | 0 | | | | | 44[13] | | 21 |
| | | | | 250 | | | | | 110 | | |
| | | | | 500 | | | | | 115 | | |
| | | | | 0 | | | | | 11 | | |
| | | | 3.5 | 250 | | | | | 23 | | |
| | | | | 500 | | | | | 25 | | |
| *Salmo trutta* | adult | | 105.4 | 105 | 45 | | 96 hr | | 18.5[14] | | 22 |
| | | | | | | | | | 23.0 | | |
| | | | | | | | | | 16.5 | | |
| | | | | | | | | | 16.0 | | |
| | | | | | | | | | 85.0 | | |
| | 1-yr | 5.2 | 5.0 | 410 | | | 64 hr | 0 | 38.2 | | 14 |
| | 2-yr | | | | | | | 35 | 32.0 | | |

| | | | | | | | | | |
|---|---|---|---|---|---|---|---|---|---|
| | adult | 13 | 5.2 | 0 | | 2.57 | 6 wk | 100 | 23 |
| | | | | 54 | | 2.57 | | 94 | |
| | | | | 0 | | 1.28 | | 100 | |
| | | | | 54 | | | | 54 | |
| | | | | 0 | | 0.32 | | 96 | |
| | | | | 54 | | 0.32 | | 30 | |
| | | | | 0 | | 16.0 | | 81 | |
| | | | | 81 | | 16.0 | | 81 | |
| | | | | 0 | | 4.0 | | 81 | |
| | | | | 81 | | 4.0 | | 10 | |
| (BB) | | 11 | 4.0 | 210 | | 0.7 | 9 d | 54 | 24 |
| (LD) | | | | | | | | 0 | |
| (BB) | | | 4.4 | | | | | 80 | |
| (LD) | | | | | | | | 45 | |
| (both) | | | 5.4 | | | | | 100 | |
| *Salvelinus fontinalis* | egg | 9.5 | 7.0 | 0 | | 2.2 | eyed | 95 | 25 |
| | | | 5.2 | | | | | 98 | |
| | | | | 175 | 160 | | | 94 | |
| | | | | 345 | 222 | | | 90 | |
| | | | 4.61 | 0 | | | | 92 | |
| | | | | 200 | 190 | | | 91 | |
| | | | | 500 | 490 | | | 100 | |
| | | | 4.4 | 0 | | | | 57 | |
| | | | | 110 | 100 | | | 85 | |
| | | | | 300 | 300 | | | 93 | |
| | | | 4.18 | 0 | | | | 1 | |
| | | | | 200 | 200 | | | 55 | |
| | | | | 510 | 505 | | | 81 | |
| | | | 7.0 | 0 | | | hatch | 94 | |

*(continued)*

Table 7. (*Continued*)

| Species | Life stage | Temp (° C) | pH | Total Al | Available Al | $Ca^{2+}$ | Exposure time | Survival (%) | $LT_{50}$ (hr) | Length (mm) Weight (mg) | Reference |
|---|---|---|---|---|---|---|---|---|---|---|---|
| | | | Experimental conditions | | | | | Factor | | | |
| | | | 5.2 | 0 | | | | 98 | | | |
| | | | | 175 | 160 | | | 92 | | | |
| | | | | 345 | 220 | | | 88 | | | |
| | | | 4.6 | 0 | | | | 92 | | | |
| | | | | 200 | 190 | | | 91 | | | |
| | | | | 500 | 490 | | | 100 | | | |
| | | | 4.4 | 0 | | | | 54 | | | |
| | | | | 110 | 100 | | | 81 | | | |
| | | | | 300 | 300 | | | 90 | | | |
| | | | 4.2 | 0 | | | | 0 | | | |
| | | | | 200 | 200 | | | 46 | | | |
| | | | | 510 | 510 | | | 69 | | | |
| | | 11 | 6.5 | 6 | | 1.48 | 28 d | 74 | | | 26 |
| | | | 5.2 | | | | | | | | |
| | | | | 54 | | | | 81 | | | |
| | egg[15] | | | 350 | | | | 75 | | | |
| | | | | 1090 | | | | 79 | | | |
| | | | 4.8 | 6 | | | | 15 | | | |
| | | | | 54 | | | | 53 | | | |
| | | | | 350 | | | | 34 | | | |
| | | | 4.4 | 6 | | | | 0 | | | |
| | | | | 54 | | | | 8 | | | |
| | | | | 350 | | | | 0 | | | |

| | | | | | | | | | | |
|---|---|---|---|---|---|---|---|---|---|---|
| | egg[16] | | 6.5 | 5 | | | 21 d | 100 | | |
| | | | 5.2 | | | | | 66 | | |
| | | | | 18 | | | | 79 | | |
| | | | | 349 | | | | 70 | | |
| | | | 4.8 | 5 | | | | 70 | | |
| | | | | 1167 | | | | 73 | | |
| | | | 4.4 | 5 | | | | 74 | | |
| | | | | 1167 | | | | 66 | | |
| | | | 4.0 | 5 | | | | 47 | | |
| | | | | 349 | | | | 61 | | |
| | | | | 1167 | | | | 81 | | |
| | | | | | | | 600 °d | 87.6 | | 14 |
| *Salvelinus fontinalis* | alevin | 9.5 | 7.0 | 0 | | 2.2 | 14 d | 100 | 71.6 | 25 |
| | | | 5.44 | | | | | 100 | 68.5 | |
| | | | 5.45 | 70 | 50 | | | 100 | 61.5 | |
| | | | 5.47 | 175 | 55 | | | 50 | 47.5 | |
| | | | 5.16 | 0 | | | | 100 | 68 | |
| | | | 5.21 | 170 | 150 | | | 88 | 50.5 | |
| | | | 5.12 | 450 | 260 | | | 0 | | |
| | | | 4.9 | 0 | | | | 100 | 63.5 | |
| | | | 4.91 | 100 | 90 | | | 97 | 54.5 | |
| | | | 4.88 | 300 | 290 | | | 97 | 50.5 | |
| | | | 4.61 | 0 | | | | 100 | 53.5 | |
| | | | 4.57 | 205 | 200 | | | 88 | 50.5 | |
| | | | 4.58 | 515 | 500 | | | 61 | 53.5 | |
| | | | 4.40 | 0 | | | | 98 | 50.0 | |
| | | | | 310 | 310 | | | 77 | 53.5 | |

(*continued*)

Table 7. (*Continued*)

| Species | Life stage | Temp (° C) | pH | Total Al | Available Al | $Ca^{2+}$ | Exposure time | Survival (%) | $LT_{50}$ (hr) | Length (mm) Weight (mg) | Reference |
|---|---|---|---|---|---|---|---|---|---|---|---|
| | | 12 | 7.2 | 415 | | 3.00 | 3 d | 100[a] | | 0.27[a] | 27 |
| | | | 6.1 | 324 | | | | 100[a] | | 0.26[a] | |
| | | | 5.3 | 252 | | | | 98[a] | | 0.25[a] | |
| | | | 7.2 | 254 | | | 28 d | 100[a] | | 0.58[b] | |
| | | | 6.1 | 212 | | | | 72[b] | | 0.20[a] | |
| | | | 5.3 | 198 | | | | 57[c] | | 0.19[a] | |
| | | | 7.2 | 268 | | 3.00 | 56 d | 99[a] | | 1.50[b] | |
| | | | 6.1 | 217 | | | | 52[b,c] | | 0.44[a] | |
| | | | 5.3 | 207 | | | | 27[c] | | 0.24[a] | |
| | | | 7.0 | 5 | | 3.30 | 30 d | 95 | | 16.4 | 28 |
| | | | 5.7 | 8 | | | | 98 | | 14.3 | |
| | | | 5.6 | 142 | | | | 96 | | 14.0 | |
| | | | | 292 | | | | 73* | | 12.2* | |
| | | | 6.9 | 5 | | | | 99 | | 22.1 | |
| | | | 6.5 | 4 | | | | 99 | | 20.4 | |
| | | | | 169 | | | | 96 | | 15.9* | |
| | | | | 350 | | | | 91* | | 14.4* | |
| | 15–30d | | 7.2 | 0 | | 3.30 | 15 d | 96.8[a] | | 19.5[a,b] | 29 |
| | | | 5.5 | | | | | 95.7[a] | | 18.7[b] | |
| | | | 4.5 | | | | | 61.4[c] | | 17.0[c] | |
| | | | 7.2 | 300 | | | | 97.2[a] | | 19.6[a] | |
| | | | 5.5 | | | | | 84.0[b] | | 17.2[c] | |
| | | | 4.5 | | | | | 55.1[c] | | 16.4[c] | |

| | | | | | | | | | | | |
|---|---|---|---|---|---|---|---|---|---|---|---|
| | alevin | | 5.2 | 20 | | | 14 d | | > 500 | | 30 |
| | | | | 420 | | | | | 115 | | |
| | | | 4.4 | 10 | | | | | > 500 | | |
| | | | | 480 | | | | | 256 | | |
| | | 6 | 4.8 | 0 | | | 13 d | 100[a] | | | 31 |
| | | | | 200 | | | | 88[a] | | | |
| | | | 4.2 | 0 | | | | 54[b] | | | |
| | | | | 100 | | | | 33[c] | | | |
| | | | 6.5 | 5 | | | 21 d | 100 | | | 26 |
| | | | 4.8 | | | | | 100 | | | |
| | | | | 1187 | | | | 100 | | | |
| | | | 4.4 | 5 | | | | 100 | | | |
| | | | | 1187 | | | | 32 | | | |
| | | | 4.0 | 5 | | | | 0 | | | |
| | | 11 | 5.5 | 51 | | 2.39 | 30 d | 70[a] | | | 32 |
| | | | | | | | | 35[b] | | | |
| | | | | | | | | 4[c] | | | |
| | | | | | | | 800 °d | 87.1 | | | 10 |
| *Salvelinus fontinalis* | juvenile | 9.5 | 7.0 | 0 | | 2.2 | 14 d | 99 | | 112 | 25 |
| | | | 5.47 | | | | | 100 | | 103.5 | |
| | | | 5.48 | 85 | 45 | | | 98 | | 88.5 | |
| | | | | 145 | 75 | | | 45 | | 70 | |
| | | | 5.20 | 0 | | | | 100 | | 106.5 | |
| | | | 5.17 | 180 | 145 | | | 68 | | 82.0 | |
| | | | 5.16 | 400 | 280 | | | 0 | | | |
| | | | 4.94 | 0 | | | | 98 | | 95.5 | |
| | | | 4.92 | 100 | 80 | | | 98 | | 80.5 | |
| | | | 4.90 | 305 | 270 | | | 39 | | 88.0 | |
| | | | 4.41 | 0 | | | | 96 | | 88.0 | |
| | | | 4.43 | 310 | 295 | | | 58 | | 69.5 | |

(*continued*)

Table 7. (*Continued*)

| Species | Life stage | Temp (° C) | pH | Total Al | Available Al | $Ca^{2+}$ | Expo-sure time | Survival (%) | $LT_{50}$ (hr) | Length (mm) Weight (mg) | Refer-ence |
|---|---|---|---|---|---|---|---|---|---|---|---|
| | | Experimental conditions | | | | | | Factor | | | |
| *Salvelinus fontinalis* | 37–67 d | 12 | 7.2 | 0 | | 3.30 | 15 d | 98.7[a] | | 28.4[a] | 29 |
| | | | 5.5 | | | | | 97.5[ab] | | 28.0[a] | |
| | | | 4.5 | | | | | 96.2[b] | | 26.5[c] | |
| | | | 7.2 | 300 | | | | 98.7[a] | | 28.4[a] | |
| | | | 5.5 | | | | | 68.0[c] | | 25.0[b] | |
| | | | 4.5 | | | | | 77.5[d] | | 25.7[b] | |
| *Salvelinus fontinalis* | juvenile | | 7.5 | 0 | | 0.09 | | 92[a] | | 31.9[a] | 33 |
| | | | | 300 | | 0.23 | | 84[a] | | 31.6[a] | |
| | | | 5.5 | 0 | | 0.13 | | 62[b] | | 26.1[b] | |
| | | | | 300 | | 0.27 | | 0[c] | | | |
| | | | 4.5 | 0 | | 0.12 | | 0[c] | | | |
| | | | | 300 | | 0.22 | | 13[c] | | | |
| | | | 7.5 | 0 | | 0.09 | 45 d | 88[a] | | 38.3[a] | |
| | | | | 300 | | 0.23 | | 80[a] | | 37.3[b] | |
| | | | 5.5 | 0 | | 0.13 | | 40[b] | | 32.8[c] | |
| | | 6 | 4.8 | 0 | | | 13 d | 100[a] | | | 31 |
| | | | | 200 | | | | 95[a] | | | |
| | | | 4.2 | 0 | | | | 67[b] | | | |
| | | | | 50 | | | | 34[c] | | | |
| | | | 6.5 | 5 | | 1.48 | 12 d | 100 | | | 26 |
| | | | 5.2 | | | | | 90 | | | |
| | | | | 349 | | | | 90 | | | |

| | | | | | | | |
|---|---|---|---|---|---|---|---|
| | 4.4 | 5 | | | 87 | | |
| | | 349 | | | 81 | | |
| | | 1167 | | | 0 | | |
| | 4.0 | 5 | | | 0 | | |
| 10.9 | 5.2 | 5 | 0.37 | 91 d | 30 | | |
| | | 18 | | | 83.1 | | |
| | | 121 | | | 79.5 | | |
| | | 349 | | | 11 | | |
| | 4.8 | 5 | | | 5 | | |
| | | 121 | | | 22.3 | | |
| | | 349 | | | 19.3 | | |
| | 4.4 | ≤1167 | | | 0 | | |
| | 5.2 | 5 | 1.48 | | 71.6 | | |
| | | 48 | | | 88.5 | | |
| | | 349 | | | 41.0 | | |
| | 4.8 | 5 | | | 57.8 | | |
| | | 349 | | | 92.7 | | |
| | 4.4 | ≤1167 | | | 0 | | |
| | 4.8 | 5 | 6.23 | | 5 | | |
| | | 1167 | | | 99 | | |
| | 4.4 | 5 | | | 0 | | |
| | | 121 | | | 29 | | |
| | | 1167 | | | 0 | | |
| | 4.0 | ≤1167 | | | 0 | | |
| | | 0 | | 14 d | | 4.5 | 34 |
| | | 1000 | | | | 5.0 | |
| | 4.4 | 0 | | | | 10.4 | |
| | | 100 | | | | 11.5 | |
| | | 500 | | | | 2.1 | |

*(continued)*

Table 7. (*Continued*)

| Species | Life stage | Experimental conditions: Temp (° C) | pH | Total Al | Available Al | $Ca^{2+}$ | Exposure time | Factor: Survival (%) | $LT_{50}$ (hr) | Length (mm) Weight (mg) | Reference |
|---|---|---|---|---|---|---|---|---|---|---|---|
| | | | 4.9 | 0 | | | | 100 | > 14 | 30.2 | |
| | | | | 100 | | | | 95 | > 14 | 28.8 | |
| | | | 4.9 | 500 | | | | | 2.3 | | |
| | | | 5.2 | 0 | | | | 93 | > 14 | 27.5 | |
| | | | | 250 | | | | 63 | > 14 | 27.3 | |
| | | | | 500 | | | | | 1.6 | | |
| | | 17 | —[17] | 200 | | 2.0 | 24 d | 92 | | | 35 |
| | | | 4.9 | | | | 8 d | 76 | | | |
| | | | —[17] | 400 | | | 24 d | 60 | | | |
| | | | 4.9 | | | | 8 d | 60 | | | |
| | | | 4.4 | | | | | 100 | | | |
| | | | | 800 | | | | 0 | | | |
| | | | 4.4 | | | | | 26 | | | |
| | | 12 | 6.5 | 0 | | | 91 d | 31 | | 49.8 | 36 |
| | | | 5.2 | | | | | 30 | | 45.2 | |
| | | | | 111 | | | | 29 | | 64.5 | |
| | | | | 333 | | | | 17 | | 52.9 | |
| | | | 4.8 | 0 | | | | 21 | | 39.2 | |
| | | | | 111 | | | | 20 | | 60.1 | |
| | | | | 333 | | | | 39 | | 49.5 | |
| | | | 4.4 | —[18] | | | | ≤4 | | | |

| | | | | | | | | | | |
|---|---|---|---|---|---|---|---|---|---|---|
| *Salvelinus fontinalis* | adult | 10 | 5.2 | 0 | | 0.50 | 10 d | 100 | | 37 |
| | | | | 111 | | | | 100 | | |
| | | | 5.2 | 333 | | | | 0 | | |
| | | | | 0 | | 8.00 | | 100 | | |
| | | | | 111 | | | | 100 | | |
| | | | | 333 | | | | 100 | | |
| | | | 4.8 | 0 | | 0.50 | | 100 | | |
| | | | | 111 | | | | 100 | | |
| | | | | 333 | | | | 17 | | |
| | | | | 0 | | 8.00 | | 100 | | |
| | | | | 111 | | | | 100 | | |
| | | | | 333 | | | | 83 | | |
| | | | 5.4 | 0 | | 1.89 | 28 d | 98 | 10.5 | 38 |
| | | | | 118 | 29 | 1.78 | | 78 | 7.8 | |
| | | | 4.8 | 4 | 4 | 1.76 | | 100 | 8.7 | |
| | | | | 191 | 237 | | | 57 | 6.0 | |
| | | | 4.4 | 5 | 4 | 1.85 | | 96 | 8.6 | |
| | | | | 665 | 552 | 1.85 | | 21 | 5.0 | |
| | | | | 8 | 7 | 1.05 | | 100 | 7.6 | |
| | | | | 296 | 239 | 0.98 | | 77 | 7.9 | |
| | | | | 704 | 443 | 0.96 | | 14 | | |
| | | 11 | 5.1 | 81 | 21 | 6.8 | 54 d | 100 | | 39 |
| | | | 5.0 | 77 | 17 | 0.5 | | 72 | | |
| | | | 5.1 | 136 | 62 | 6.8 | | 100 | | |
| | | | | 156 | 48 | 0.5 | | 84 | | |
| | | 5.2 | 5.0 | 415 | | | 64 hr | 80 | | 14 |
| | | 10 | 4.8 | 0 | | 0.5 | 256 hr | 100 | | 40 |
| | | | | 333 | | —[19] | | 0 | | |
| | | | 4.4 | 0 | | 0.50 | | 41 | | |
| | | | | 333 | | | | 15 | | |

*(continued)*

Table 7. (*Continued*)

| Species | Life stage | Experimental conditions | | | | | | Factor | | | Reference |
|---|---|---|---|---|---|---|---|---|---|---|---|
| | | Temp (° C) | pH | Total Al | Available Al | $Ca^{2+}$ | Exposure time | Survival (%) | $LT_{50}$ (hr) | Length (mm) Weight (mg) | |
| | | 11 | 5.2 | —[20] | | 8.0 | 10 wk | 0 | | | 41 |
| | | | | 75 | | 0.5 | | 28 | | | |
| | | | | 150 | | | | 16 | | | |
| *Salvelinus namaycush* | alevin | 5 | 6.9 | 4 | | 1.54 | 5 d | | | 27.1[a,21] | 42 |
| | | | 5.1 | 5 | | 1.58 | | | | 27.0[a] | |
| | | | 5.0 | 101 | | 1.53 | | | | 26.8[a] | |
| | | | 5.1 | 199 | | 1.63 | | | | 26.0[b] | |
| | | | 6.9 | 4 | | 1.54 | | | | 27.2[a] | |
| | | | 5.1 | 5 | | 1.58 | | | | 27.2[a] | |
| | | | 5.0 | 101 | | 1.53 | | | | 26.3[b] | |
| | | | 5.1 | 199 | | 1.63 | | | | 26.2[b] | |
| | | 6 | 4.8 | 0 | | | 13 d | 99[a] | | | 31 |
| | | | | 300 | | | | 70[b] | | | |
| | | | 4.6 | 0 | | | | 100[a] | | | |
| | | | | 50 | | | | 52[c] | | | |
| | | | 4.4 | 0 | | | | 18[d] | | | |
| | | | | 50 | | | | 6[e] | | | |
| | | | 4.2 | ≥0 | | | | 0[e] | | | |
| *Salvelinus namaycush* | juvenile | | 4.8 | 0 | | | | 99[a] | | | |
| | | | | 300 | | | | 82[b] | | | |
| | | | 4.6 | 0 | | | | 100[a] | | | |
| | | | | 300 | | | | 71[b] | | | |

| | | | | | | | | | | |
|---|---|---|---|---|---|---|---|---|---|---|
| | | 6 | 4.4 | 0 | | | | 43[c] | | |
| | | | | 50 | | | | 27[d] | | |
| | | | 4.2 | ≥0 | | | | 3[e] | | |
| *Pimephales promelas* | eggs | 25 | 7.5 | 22 | 14 | | hatch | 92[a] | | 43 |
| | | | | 66 | 35 | | | 94[a] | | |
| | | | 6.0 | 25 | 14 | | | 60[b] | | |
| | | | 5.5 | 24 | 15 | | | 0[c] | | |
| | | | | 49 | 28 | | | 38[b] | | |
| *Pimephales promelas* | alevin | | 7.5 | 22 | 14 | | —[22] | 92[a] | | |
| | | | | 66 | 35 | | | 92[a] | | |
| | | | 6.0 | 25 | 14 | | | 56[b] | | |
| | | | 5.5 | 24 | 15 | | | 0[c] | | |
| | | | 7.5 | 22 | 14 | | 14 d[23] | 98[a] | 0.65[a,24] | |
| | | | | 66 | 35 | | | 100[a] | 0.82[a] | |
| | | | 6.0 | 25 | 14 | | | 73[b] | 0.74[a] | |
| *Pimephales promelas* | juvenile | 21 | 6.5 | 0 | | 6.9 | 96 hr | 100 | | 44 |
| | | | | 28 | 16 | | | 100 | | |
| | | | 5.5 | 0 | | | | 100 | | |
| | | | | 26 | 30 | | | 70 | | |
| | | | | 41 | 51 | | | 0 | | |
| | | | 4.5 | 0 | | | | 0 | | |
| *Catostomus commersoni* | egg | 18 | 7.1 | 0 | | 2.2 | eyed | 78.3 | | 25 |
| | | | 5.58 | 0 | | | | 51.4 | | |
| | | | 5.59 | 70 | 25 | | | 4.1 | | |
| | | | 5.17 | 0 | | | | 30.9 | | |
| | | | 5.18 | 155 | 135 | | | 32.7 | | |

*(continued)*

Table 7. (*Continued*)

| Species | Life stage | Experimental conditions: Temp (° C) | pH | Total Al | Available Al | $Ca^{2+}$ | Exposure time | Factor: Survival (%) | $LT_{50}$ (hr) | Length (mm) Weight (mg) | Reference |
|---|---|---|---|---|---|---|---|---|---|---|---|
| | | | 5.17 | 350 | 175 | | | 21.0 | | | |
| | | | 4.84 | 0 | | | | 0 | | | |
| | | | 4.77 | 190 | 175 | | | 75.3 | | | |
| | | | | 475 | 460 | | | 0 | | | |
| | | | 7.1 | 0 | | | hatch | 76.9 | | | |
| | | | 5.58 | | | | | 51.4 | | | |
| | | | 5.59 | 70 | 25 | | | 4.1 | | | |
| | | | 5.17 | 0 | | | | 29.8 | | | |
| | | | 5.18 | 155 | 135 | | | 0 | | | |
| | | 4 | 6.0 | 0 | | 4.0 | eyed | 90.9 | | | 3 |
| | | | | 50 | | | | 89.9 | | | |
| | | | 5.4 | 100 | | | | 90.7 | | | |
| | | | 5.1 | 0 | | | | 80.1 | | | |
| | | | 5.1 | 200 | | | | 98.0 | | | |
| | | | 4.8 | 0 | | | | 72.4 | | | |
| | | | | 50 | | | | 87.5 | | | |
| | | | | 200 | | | | 76.6 | | | |
| | | | 4.5 | 0 | | | | 1.0 | | | |
| | | | | 50 | | | | 43.8 | | | |

| | | | | | | | | | | |
|---|---|---|---|---|---|---|---|---|---|---|
| *Catostomus commersoni* | alevin | 18 | 7.1 | 0 | | 2.2 | 14 d | 86.7 | 20.7 | 25 |
| | | | 5.59 | 0 | | | | 98.4 | 17.5 | |
| | | | 5.60 | 105 | 40 | | | 5.0 | 5.0 | |
| | | | 5.58 | 305 | 55 | | | 0 | | |
| | | | 5.38 | 0 | | | | 95.3 | 15.5 | |
| | | | 5.41 | 45 | 50 | | | 0 | | |
| | | | 5.40 | 155 | 75 | | | 46.7 | 4.0 | |
| | | | 5.22 | 0 | | | | 98.4 | 15.5 | |
| | | | 5.18 | 170 | 125 | | | 0 | | |
| | | | 5.02 | 0 | | | | 48.4 | 8.5 | |
| | | | | 85 | 85 | | | 10.3 | 4.0 | |
| | | 13 | 5.4 | 0 | | 4.0 | 18 d[25] | 75.5[a] | | 3 |
| | | | | 100 | | | | 42.4[c] | | |
| | | | 5.1 | 0 | | | | 60.2[b] | | |
| | | | | 50 | | | | 76.5[a] | | |
| | | | | 100 | | | | 50.1[c] | | |
| | | | 4.8 | 0 | | | | 64.7[b] | | |
| | | | | 50 | | | | 60.6[b] | | |
| | | | | 100 | | | | 6.3[d] | | |
| *Catostomus commersoni* | juvenile | 18 | 6.74 | 0 | | 2.2 | 14 d | 95.1 | 53.3 | 25 |
| | | | 5.55 | 0 | | | | 97.4 | 49.0 | |
| | | | 5.58 | 85 | 45 | | | 71.6 | 35.0 | |
| | | | 5.57 | 305 | 40 | | | 5.0 | 17.0 | |
| | | | 5.21 | 10 | | | | 97.5 | 42.0 | |

(*continued*)

Table 7. (*Continued*)

| Species | Life stage | Temp (° C) | pH | Total Al | Available Al | $Ca^{2+}$ | Exposure time | Survival (%) | $LT_{50}$ (hr) | Length (mm) Weight (mg) | Reference |
|---|---|---|---|---|---|---|---|---|---|---|---|
| | | | Experimental conditions | | | | | Factor | | | |
| | | | 5.22 | 120 | 80 | | | 39.0 | | 14.0 | |
| | | | 5.17 | 350 | 170 | | | 0 | | | |
| | | | 4.98 | 10 | | | | 93.2 | | 42 | |
| | | | 5.00 | 90 | 75 | | | 42.8 | | 20 | |
| | | | | 220 | 220 | | | 0 | | | |
| | | | 4.80 | 0 | | | | 87.5 | | 30.0 | |
| | | | 4.82 | 195 | 155 | | | 0 | | | |
| | juvenile[26] | 18 | 5.0 | 60 | | | 5 d | | 140 | | 30 |
| | | | | 221 | | | | | 37 | | |
| | | | | 383 | | | | | 17 | | |
| | juvenile[27] | | | 100 | | | | | 500 | | |
| | | | | 200 | | | | | 27 | | |
| | | | | 291 | | | | | 22 | | |
| | | | | 500 | | | | | 13 | | |
| *Morone saxatilis* | alevin[28] | 14.5 | 7.2 | 150 | | | 96 hr | 84 | | | 45 |
| | | 14.3 | 6.4 | 2550 | | | | 1 | | | |
| | | 14.5 | 6.3 | 1250 | | | 96 hr | 5 | | | |
| | | | | 750 | | | | 9.5 | | | |
| *Morone saxatilis* | juvenile[29] | 18 | 7.2 | 0 | | | 7 d | 74 | | | 46 |
| | | | | 200 | | | | 74 | | | |
| | | | | 400 | | | | 4* | | | |

| | | | | | | | | | |
|---|---|---|---|---|---|---|---|---|---|
| | | | 6.5 | 0 | | | | 79 | |
| | | | | 25 | | | | 42* | |
| | | | 6.0 | 0 | | | | 2 | |
| | | | | 25 | | | | 6 | |
| | juvenile[30] | | 7.2 | 0 | | | | 100 | |
| | | | | 400 | | | | 0* | |
| | | | 6.5 | 0 | | | | 100 | |
| | | | | 200 | | | | 42* | |
| | | | 6.0 | 0 | | | | 98 | |
| | | | | 50 | | | | 62* | |
| *Micropterus dolomieui* | alevin[31] | 18 | 6.0 | 0 | 0 | 4.0 | 7 d | 97.3[a] | 3 |
| | | | 4.8 | 0 | 2 | | | 98.0[a] | |
| | | | | 300 | 284 | | | 98.0[a] | |
| | | | 4.5 | 0 | 2 | | | 61.8[b] | |
| | | | | 100 | 67 | | | 98.0[a] | |
| | | | | 300 | 270 | | | 93.2[a] | |
| | | | 4.2 | 0 | 5 | | | 45.3[c] | |
| | | | | 100 | 81 | | | 67.4[b] | |
| | | | | 300 | 287 | | | 77.9[b] | |
| | alevin[32] | | 6.0 | 0 | 0 | | 11 d | 97.3[a] | |
| | | | 4.8 | 0 | 2 | | | 97.3 | |
| | | | | 100 | 81 | | | 96.7[a] | |
| | | | | 200 | 192 | | | 50.0[c] | |
| | | | 4.5 | 0 | 2 | | | 74.7[b] | |
| | | | | 100 | 81 | | | 53.1[c] | |
| | | 18 | 4.2 | 0 | 5 | 4.0 | 11 d | 0.0[e] | 3 |
| | | | | 100 | 81 | | | 6.8[d] | |

*(continued)*

Table 7. (*Continued*)

| | | Experimental conditions | | | | | | Factor | | | |
|---|---|---|---|---|---|---|---|---|---|---|---|
| Species | Life stage | Temp (° C) | pH | Total Al | Available Al | $Ca^{2+}$ | Exposure time | Survival (%) | $LT_{50}$ (hr) | Length (mm) Weight (mg) | Reference |
| | alevin | 21 | 7.5 | 2.3 | | 3.1 | 96 hr | 95 | | | 47 |
| | | | | 100 | | | | 90 | | | |
| | | | | 217 | 11 | | | 80 | | | |
| | | | 6.1 | 0.3 | | | | 95 | | | |
| | | | | 978 | 11 | | | 95 | | | |
| | | | 5.1 | 2.3 | | | | 95 | | | |
| | | | | 219 | 54 | | | 0 | | | |
| | | 20 | 7.3 | 1.3 | | 3.0 | 30 d | 63 | | 13.8 | |
| | | | | 252 | | | | 51 | | | |
| | | | 5.7 | 0.5 | | | | 54 | | 14.6 | |
| | | | 5.5 | 205 | | | | 0 | | 15.5 | |
| | | | 5.1 | 2.3 | | | | 28 | | | |
| | | | | 218 | | | | 0 | | | |
| *Stizostedion vitreum* | egg | 11 | 6.0 | 0 | | 4.0 | hatch[33] | 67.9[a] | | | 3 |
| | | | 5.4 | 0 | | | | 61.1[a] | | | |
| | | | 5.4 | 100 | | | | 42.8[b] | | | |
| | | | 4.9 | 0 | | | | 46.3[b] | | | |
| | | | | 50 | | | | 8.3[c] | | | |
| | | | 4.4 | 0 | | | | 0[d] | | | |
| *Stizostedion vitreum* | alevin | 19.8 | 6.0 | | | | hatch[34] | 89.9[a] | | | |
| | | | | 50 | | | | 94.4[a] | | | |

| | | |
|---|---|---|
| 5.0 | 0 | 80.9[a] |
| | 100 | 100[a] |
| | 200 | 18.4[c] |
| 4.7 | 0 | 80.9[a] |
| | 100 | |
| | 200 | 66.7[b] |
| 4.5 | 0 | 57.1[b] |
| | 200 | 80.9[a] |

1. Embryos 20 to 24 hr old, early tail-bud stage.
2. Represents median time for exposures ranging from 66 to 72 hr.
3. Duration of a pulse exposure followed by exposure of pH 7.5 and 0 $\mu g \cdot L^{-1}$ Al.
4. Results with different letters used as superscripts are significantly different.
5. Asterisks identify results that are significantly different from results observed at the same pH and 0 $\mu g \cdot L^{-1}$ Al; plus (+) identifies results that are significantly different from a circumneutral control and 0 $\mu g$ $Al \cdot L^{-1}$.
6. Challenge test result after acclimation to pH 5.2 water for 21 d.
7. Challenge test result after acclimation to pH 5.2, 27 $\mu g \cdot L^{-1}$ Al for 21 d.
8. Calcium levels 1.8 and 16.4 $mg \cdot L^{-1}$.
9. Parr 1+ years in age.
10. Parr 2+ years in age, smoltification underway.
11. Treatments both with and without aluminum.
12. Total period of exposure from egg fertilization through alevin stage.
13. Median survival time.
14. Results reflect additions of 0, 2.9, 5.8, 11.5, and 92.0 $mg \cdot L^{-1}$ of sodium, respectively (Dietrich et al. 1989).
15. Exposures began with egg fertilization.
16. Exposures began with eyed embryos.
17. Represents treatments at pH 4.4 and 4.9 together.
18. Al concentrations 0, 12, 37, 111, 333, and 1000 $\mu g \cdot L^{-1}$.
19. Represents results of fish exposed to 0.5 and 8.0 $mg \cdot L^{-1}$ Ca.

(*continued*)

Table 7. (*Continued*)

20. Represents results of fish exposed to 75 and 150 $\mu g \cdot L^{-1}$ Al.
21. Letters apply to differences among treatments of alevins that experienced the same incubation and recovery conditions.
22. Exposed from egg stage through 4 d after hatching.
23. Exposed for 14 d starting within 24 hr after hatching.
24. Mean dry weight per individual 14-d alevin.
25. Data taken 4 d posthatch.
26. Length 14 mm.
27. Length 19 mm.
28. Juveniles 11 d old at start of 7-d test.
29. Juveniles 160 d old at start of 7-d test.
30. Exposures began 24 hr posthatch.
31. Exposures began 3 d before hatching and continued until 4 d posthatch.
32. Exposures began 3 d before hatching and continued until 8 d posthatch.
33. Hatching occurred after 16 d exposure.
34. Measurements made 4 d posthatch.

References: (1) Klauda et al. 1987; (2) Klauda and Palmer 1987; (3) Holtze and Hutchinson 1989; (4) Woodward et al. 1989; (5) Neville and Campbell 1988; (6) Reid et al. 1991; (7) Goss and Wood 1988; (8) Heming and Blumhagen 1988; (9) Playle et al. 1989; (10) Skogheim and Rosseland 1984; (11) Birchall et al. 1989; (12) Fivelstad and Leivestad 1984; (13) McKee et al. 1989; (14) Rosseland and Skogheim 1986; (15) Skogheim et al. 1986; (16) Skogheim and Rosseland 1986; (17) Brown 1983; (18) Sayer et al. 1991b; (19) Reader et al. 1991; (20) Stoner et al. 1984; (21) Brown 1981; (22) Dietrich et al. 1989; (23) Sadler and Lynam 1988; (24) Sadler and Turnpenny 1986; (25) Baker and Schofield 1982; (26) Ingersoll et al. 1990a; (27) Cleveland et al. 1991a; (28) Cleveland et al. 1989; (29) Cleveland et al. 1986; (30) Driscoll et al. 1980; (31) Hutchinson et al. 1989; (32) Mount et al. 1990; (33) Hunn et al. 1987; (34) Schofield and Trojnar 1980; (35) Siddens et al. 1986; (36) Wood et al. 1990a; (37) Booth et al. 1988; (38) Ingersoll et al. 1990b; (39) Mount et al. 1988; (40) Wood et al. 1988a; (41) Wood et al. 1988b; (42) Gunn and Noakes 1987; (43) McCormick et al. 1989; (44) Palmer et al. 1988; (45) Hall et al. 1985; (46) Buckler et al. 1987; (47) Kane and Rabeni 1987.

Table 8. Concentrations of Na, K, Mg, and Cl in plasma ($\mu g \cdot L^{-1}$) and whole bodies ($\mu g \cdot g^{-1}$) of fish exposed to different concentrations of Al ($\mu g \cdot L^{-1}$) and Ca ($mg \cdot L^{-1}$) and pH in laboratory experiments. Blanks for species name, life stage, and experimental conditions have the same values as the cell above the blank.

| | Experimental conditions | | | | | | | Ion concentrations | | | | |
|---|---|---|---|---|---|---|---|---|---|---|---|---|
| Species | Life stage | Temp (°C) | pH | Total Al | Available Al | $Ca^{2+}$ | Exposure time | $Na^+$ | $K^+$ | $Mg^{2+}$ | $Cl^-$ | Reference |
| *Oncorhynchus clarki* | alevin | 10 | 6.5 | 0 | | 1.4 | 7 d | 1334 | | | | 1 |
| | | | 6.0 | | | | | 1219 | | | | |
| | | | | 50 | | | | $667^{+*1}$ | | | | |
| | | | 5.5 | 0 | | | | $1173^+$ | | | | |
| | | | | 50 | | | | $1081^+$ | | | | |
| | | | | 100 | | | | $865^{+*}$ | | | | |
| | | | 5.0 | 0 | | | | $943^+$ | | | | |
| | | | | 50 | | | | $1035^+$ | | | | |
| *Oncorhynchus mykiss* | juvenile | | 6.5–5.5 | 0 | | 4.9–2.3 | 6–11 d | 3291 | | | 4658 | 2 |
| | | | 6.5 | 75 | | 4.9 | 11 d | 3516 | | | 4584 | |
| | | | 6.1 | | | 2.8–5.7 | 2–11 d | 3098 | | | 4064* | |
| | | | 5.0 | 0 | | 2.4–3.4 | 11 d | 3147 | | | 4435 | |
| | | | | 75 | | 2.1–3.4 | 4–11 d | $3002^+$ | | | $4411^{*+}$ | |
| | | | 4.0 | 0 | | 2.8–3.7 | 1–7 d | $2633^+$ | | | $2924^+$ | |
| | | | | 75 | | 2.3–3.7 | 5–11 d | $2472^+$ | | | $2949^+$ | |
| | | 8–12 | 5.3 | 50 | | 2.69 | 24 hr | 3500 | | | | 3 |
| | | | | 480 | | | | 2500 | | | | |

(*continued*)

Table 8. (*Continued*)

| | Experimental conditions | | | | | | | Ion concentrations | | | | |
|---|---|---|---|---|---|---|---|---|---|---|---|---|
| Species | Life stage | Temp (°C) | pH | Total Al | Available Al | $Ca^{2+}$ | Exposure time | $Na^+$ | $K^+$ | $Mg^{2+}$ | $Cl^-$ | Reference |
| | | 15 | 6.5 | 0 | | 0.65 | 8 d | | | | 4325 | 4 |
| | | | 5.2 | | | 0.68 | | | | | 4137 | |
| | | | | 27 | | 0.55 | | | | | 3420* | |
| | | | 6.5 | 0 | | 0.65 | 21 d | | | | 4340 | |
| | | | 5.2 | | | 0.68 | | | | | 4637 | |
| | | | | 27 | | 0.54 | | | | | 3695* | |
| | | | | 118 | | 0.60 | 40 hr[2] | | | | 3125 | |
| | | | | | | | | | | | 2968 | |
| | | | | | | | | | | | 3008* | |
| *Onchorhynchus mykiss* | adult | 10 | 5.2 | 0 | 0 | | 96 hr | 143 | | | 124 | 5 |
| | | | | 137 | 92 | | | 132 | | | 124 | |
| | | | | 203 | 175 | | 63 hr | 113 | | | | |
| | | | 5.4 | 0 | 0 | | 96 hr | 158 | | | 122 | |
| | | | | 142 | 123 | | | 117 | | | | |
| | | | | 490 | 193 | | 55 hr | 139 | | | 108 | |
| | | | 5.6 | 0 | 0 | | 96 hr | 149 | | | 123 | |
| | | | | 173 | 63 | | | 145 | | | 118 | |
| | | | | 530 | 341 | | 52 hr | 142 | | | 117 | |
| | | 14.5 | 6.5 | 0 | | | 72 hr | 136 | | | 130 | 6 |
| | | | 4.8 | | | | | 132 | | | 128 | |
| | | | | 112 | | | 27 hr | 130 | | | 118 | |
| | | 12.6 | 7.8 | 0 | | 38 | 0 | 152 | 3.8 | 0.48 | 137 | 7 |
| | | | 4.3 | | | | 24 hr | 145* | | | 137 | |
| | | | | | | | 96 hr | 141* | | | 129* | |

| | | | | | | | | | | |
|---|---|---|---|---|---|---|---|---|---|---|
| | | 500 | | | 24 hr | 146* | | | 137 | |
| | | | | | 96 hr | 132* | 1.9* | | 118* | |
| | 8.7 | 0 | | | | 127* | | | 118* | |
| | | 500 | | | | 153 | 2.7 | 0.31 | 133 | |
| 15 | 6.5 | 5 | | 0.9 | 0 | 3174 | 109 | | 4402 | 8 |
| | 5.2 | | | | ≤66 hr | 3013 | 106 | | 4402 | |
| | | 105 | | | | $2875^+$ | 113 | | $3728^+$ | |
| | | 5 | | 8.2 | | 3174 | 109 | | 4260 | |
| | | 105 | | | | $2783^+$ | $149^+$ | | $3870^+$ | |
| | 4.8 | 5 | | 0.9 | | 3128 | 94 | | 4402 | |
| | | 105 | | | | $2967^+$ | $125^+$ | | $3905^+$ | |
| | | 5 | | 8.2 | | 3036 | 113 | | 4260 | |
| | | 105 | | | | $2806^+$ | 109 | | $3905^+$ | |
| | 4.4 | 5 | | 0.9 | | 2921 | 106 | | $3515^+$ | |
| | | 105 | | | | $2622^+$ | 125 | | $3515^+$ | |
| | | 5 | | 8.2 | | $2852^+$ | 117 | | $4012^+$ | |
| | | 105 | | | | $2530^+$ | 121 | | $3941^+$ | |
| 14 | 5.0 | 0 | | | 0 hr | 3036 | | | | 9 |
| | | | | | 28 hr | | | | | |
| | | | | | 52 hr | 2944 | | | | |
| | | 60 | | | 28 hr | 2802* | | | | |
| | | | | | 52 hr | 2708* | | | | |
| 13 | 5.0 | 0 | | | 10 d | 2599 | | | 4615 | 10 |
| | | 200 | 190 | | 4 d | 2190* | | | 3092* | |
| 17 | 6.1 | 0 | | 0.68 | 3.5 hr | 3317 | 90 | | 4864 | 11 |
| | 6.0 | | | 3.76 | | 3289 | 106 | | 4828 | |
| | 4.1 | | | 0.80 | | $3082^+$ | 63 | | $4367^+$ | |
| | | 351 | | | | $2921^+$ | $122^+$ | | $4260^+$ | |
| | | | | 3.70 | | $3128^+$ | $173^+$ | | $4331^+$ | |

*(continued)*

Table 8. (*Continued*)

| | Experimental conditions | | | | | | | Ion concentrations | | | | |
|---|---|---|---|---|---|---|---|---|---|---|---|---|
| Species | Life stage | Temp (°C) | pH | Total Al | Available Al | $Ca^{2+}$ | Exposure time | $Na^{+}$ | $K^{+}$ | $Mg^{2+}$ | $Cl^{-}$ | Reference |
| *Salmo salar* | egg | 10 | 7.5 | 0 | 0 | | 18 d | 447 | | | | 12 |
| | | | 4.0 | | | | | 359 | | | | |
| | | | 4.6 | 1000 | | | | 308 | | | | |
| *Salmo trutta* | juvenile | 8–12 | 4.8 | 100 | | 2.69 | 24 hr | 4000 | | | | 3 |
| | | | | 480 | | | | 3100 | | | | |
| | | 10 | 4.5 | 0 | | | 30 hr | 3243 | | | 4331 | 13 |
| | | | | 323 | | | | 2268* | | | 3053* | |
| | | | | 0 | | | | 3243 | | | 4296 | |
| | | | | 323 | | | | 2668* | | | 3302* | |
| | | 10 | 4.5 | 0 | | | 96 hr | 2870 | | | | 14 |
| | | | | 324 | | | | 2093* | | | | |
| *Salvelinus fontinalis* | alevin | 12 | 7.2 | 0 | | | 160 hr | 5200[a] | 13500[3] | 1450 | | 15 |
| | | | | 300 | | | | 9800[c] | 18300[3] | 1550 | | |
| | | | 6.5 | 0 | | | | 3000[b] | 8000 | 1350 | | |
| | | | | 315 | | | | 7800[d] | 13000 | 1250 | | |
| | | | 6.0 | 0 | | | | 3100[b] | 8000 | 1330 | | |
| | | | | 373 | 124 | | | 7000[ef] | 12000 | 1270 | | |
| | | | 5.3 | 0 | | | | 3000[b] | 9000 | 1300 | | |
| | | | | 403 | 265 | | | 7000 | 10500 | 1250 | | |
| | | | 4.5 | 0 | | | | 2800[b] | 7500 | 1250 | | |
| | | | | 405 | 447 | | | 7500[de] | 12300 | 1300 | | |
| | | | 4.0 | 453 | 363 | | | 6000[f] | 10300 | 1280 | | |

| | | | | | | | | | | | |
|---|---|---|---|---|---|---|---|---|---|---|---|
| *Salvelinus fontinalis* | juvenile | 8–12 | 4.9 | 40 | 2.69 | 24 hr | 4400[3] | | | | 3 |
| | | | | 450 | | | 2910 | | | | |
| | | 9.1 | 6.6 | 0.9 | 0.56 | 4 d | 44.6[3] | | | 34.5 | 16 |
| | | 9.2 | 5.1 | 1.4 | 0.52 | | 42.6 | | | 32.0 | |
| | | 9.0 | 5.3 | 165 | 0.44 | | 35.1* | | | 24.4* | |
| | | 9.1 | 6.6 | 0.9 | 0.56 | 24 d | 44.6 | | | 34.5 | |
| | | 9.2 | 5.1 | 1.4 | 0.52 | | 39.6* | | | 31.0 | |
| | | | 5.2 | 68.2 | 0.54 | | 38.6* | | | 26.8* | |
| *Salvelinus fontinalis* | adult | 10 | 6.5 | 0 | 1.00 | 24 hr | 3450 | 127.0 | | 5133 | 17 |
| | | | 4.8 | 111 | | | 2898 | 144.6 | | 3135 | |
| | | | | 333 | | | 2760 | 127.5 | | 3387 | |
| | | | | 0 | | 11 d | 3267 | 109.5 | | 4118 | |
| | | | | 111 | | | 3224 | 98.9 | | 3135 | |
| | | | | 0 | 0.50 | 0 | 3145 | | | 4371 | 18 |
| | | | | 333 | | 42 hr | 2623* | | | 3767* | |
| | | | | | 9.20 | | 2916* | | | 4049* | |
| | | | 4.4 | 0 | 0.50 | 0 | 3109 | | | 4411 | |
| | | | | | | 42 hr | 2780* | | | 3888* | |
| | | | | 333 | | | 2675* | | | 3847* | |
| | | 11 | 6.5 | 0 | 8.00 | 10 wk | 3395 | 104 | 102[4] | 4684 | 19 |
| | | | 5.2 | | | | 3202* | 100 | 90 | 4627 | |
| | | | | 75 | | | 3670* | 123 | 133* | 4778 | |
| | | | | 150 | | | 3364 | 129* | 117 | 4613 | |
| | | | 6.5 | 0 | 0.50 | | 3181 | 89 | 111 | 4623 | |
| | | | 5.2 | | | | 3242 | 122* | 96 | 4731 | |
| | | | | 75 | | | 3052* | 113 | 107 | 4528 | |
| | | | | 150 | | | 3257 | 121* | 104 | 4764 | |

(*continued*)

Table 8. (*Continued*)

| | | Experimental conditions | | | | | | Ion concentrations | | | | |
|---|---|---|---|---|---|---|---|---|---|---|---|---|
| Species | Life stage | Temp (°C) | pH | Total Al | Available Al | $Ca^{2+}$ | Exposure time | $Na^+$ | $K^+$ | $Mg^{2+}$ | $Cl^-$ | Reference |
| *Salmo namaycush* | alevin[5] | 5 | 6.9 | 4 | | 1.54 | 5 d | 7.3[a,b] | 16.0[a] | 5.5[a] | 9.0[a] | 20 |
| | | | 5.1 | 5 | | 1.58 | | 7.5[a] | 15.7[a] | 5.4[a] | 9.0[a] | |
| | | | 5.0 | 101 | | 1.53 | | 6.7[c] | 12.7[b] | 4.5[b] | 7.9[b] | |
| | | | 5.1 | 199 | | 1.63 | | 6.8[b,c] | 13.1[b] | 4.3[b] | 7.9[b] | |
| | alevin[6] | | 6.9 | 4 | | 1.54 | | 7.0[a] | 16.0[a] | 5.9[a] | 9.0[a] | |
| | | | 5.1 | 5 | | 1.58 | | 7.2[a] | 15.2[a,b] | 5.7[a,b] | 9.0[a] | |
| | | | 5.0 | 101 | | 1.53 | | 7.3[a] | 14.4[b] | 5.4[b] | 9.1[a] | |
| | | | 5.1 | 199 | | 1.63 | | 6.5[b] | 13.0[c] | 4.7[c] | 8.0[b] | |

1. Results that are significantly different from a circumneutral control are indicated by +; those that are significantly different from fish exposed to the same pH and 0 $\mu g \cdot L^{-1}$ Al are indicated by *.

2. Challenged to 118 $\mu g \cdot L^{-1}$ Al for 40 hr after acclimation to above conditions for 21 d.

3. Whole-body concentrations. Results with different letters are significantly different.

4. Calcium ($Ca^{+2}$) levels.

5. Alevins from eggs incubated 599 °C degree-days before hatching; electrolyte measurements were taken following a 32-d recovery period.

6. Alevins from eggs incubated 712 °C degree-days before hatching; electrolyte measurements were taken following a 21-d recovery period.

References: (1) Woodward et al. 1989; (2) Neville 1985; (3) Gagen and Sharpe 1987; (4) Reid et al. 1991; (5) Dietrich and Schlatter 1989b; (6) Goss and Wood 1988; (7) Heming and Blumhagen 1988; (8) Playle et al. 1989; (9) Witters et al. 1991; (10) Witters et al. 1990a; (11) Witters 1986; (12) Eddy and Talbot 1985; (13) Reader et al. 1991; (14) Sayer et al. 1991a; (15) Cleveland et al. 1991b; (16) McDonald et al. 1991; (17) Booth et al. 1988; (18) Wood et al. 1988a; (19) Wood et al. 1988b; (20) Gunn and Noakes 1987.

Cutthroat alevins may respond according to the asphyxiation/electrolyte regulation model proposed by Neville and Campbell (1988). Mortalities and Na losses were greater at pH 6.0 and 5.0 with 50 $\mu g \cdot L^{-1}$ Al than at pH 5.5. Although mortality at pH 5.5 and 100 $\mu g \cdot L^{-1}$ Al was significantly higher than with no Al, it was about one-half that at the other two pH levels. Electrolyte regulation in alevins did not follow the model as well as did mortality. Whole-body Na was reduced significantly in alevins exposed to pH 6.0 with 50 $\mu g \cdot L^{-1}$ Al and to pH 5.5 with 100 $\mu g \cdot L^{-1}$ compared to alevins exposed to the same pH but no Al. Sodium concentrations at pH 5.0 with and without Al were not significantly different. The metal exceeded theoretical saturation levels at both pH 6.0 and 5.5 (Skogheim et al. 1987). The Na losses at pH 6.0 may have been an indirect result from asphyxiation rather than a direct loss of control of gill membrane permeability. The data for electrolyte regulation suggest that Al may have provided some protection against the high $H^+$ present at pH 5.0, although differences were not significant.

Although rainbow trout has been one of the most commonly used species in studies of Al effects on fish, almost all studies have focused on juveniles and adults (e.g., Neville 1985; Neville and Campbell 1988; Playle et al. 1989; Playle and Wood 1991). Both cutthroat and rainbow trout respond similarly to Al exposures.

Results of some studies of juvenile rainbow trout suggest that prior exposure to very low Al levels may increase resistance of younger trout. For instance, the $LT_{50}$ of juveniles that were exposed to pH 5.2 and no Al for 21 d and then to the same pH and 118 $\mu g \cdot L^{-1}$ was 31.5 hr, but it increased to 70 hr when juveniles were first exposed to pH 5.2 and 27 $\mu g \cdot L^{-1}$ Al for 21 d (Table 7). The hematocrit of Al-exposed fish was higher than that of unexposed fish after 8 d of exposure, but dropped back to normal after 21 d of exposure (Table 9). After 8 d, plasma Cl levels were significantly lower in Al-exposed fish than in unexposed fish and remained significantly lower through the end of the exposure period (Table 8).

Adult rainbow trout appear to be as sensitive as juveniles to dissolved Al at pH ~4–5. Plasma Na and Cl were reduced after exposure to 100–150 $\mu g \cdot L^{-1}$ Al (Dietrich and Schlatter 1989b; Playle et al. 1989), and plasma Na may be significantly reduced with as little as 60 $\mu g \cdot L^{-1}$ at pH 5 (Witters et al. 1991) (Table 8). Significant differences in hematocrit counts, plasma pH, and $O_2$ pressure generally occurred at these pH and Al levels (Table 9). Adult mortality after a few days of exposure to 50–100 $\mu g \cdot L^{-1}$ Al and pH 4.3–5.2 was higher than at comparable pH without Al (Table 7) (Goss and Wood 1988; Heming and Blumhagen 1988; Playle et al. 1989).

Aluminum concentrations in gills of laboratory-exposed rainbow trout generally followed the pattern described by Neville and Campbell (1988) (Table 10). However, we found no field data that corroborated the laboratory information. The most comprehensive set of field data are of tissues of

Table 9. Changes in hematocrit values (Hct) (%), plasma pH, oxygen tension ($PO_2$) (mm Hg), and plasma lactate concentration ($\mu eq \cdot L^{-1}$) in fish exposed to different concentrations of Al ($\mu g \cdot L^{-1}$), Ca ($mg \cdot L^{-1}$), and pH in laboratory experiments. Species name, life stage, and experimental condition cells with missing values have the same values as the preceding cell within a study.

| | | Experimental conditions | | | | | | Blood chemistry | | | | |
|---|---|---|---|---|---|---|---|---|---|---|---|---|
| Species | Life stage | Temp (°C) | pH | Total Al | Available Al | $Ca^{2+}$ | Exposure time | Hct (%) | pH | $PO_2$ | Lactate | Reference |
| *Oncorhynchus mykiss* | juvenile | 10 | 6.5–5.5 | 0 | | 4.9–2.3 | 11 d | | 12.2 | 88 | 0.5 | 1 |
| | | | 6.5 | 75 | | 4.9 | | | 13.0 | 91 | 0.5 | |
| | | | 6.1 | | | 2.8–5.7 | 3–11 d | | 16.4*[1] | 41* | 3.75* | |
| | | | 5.0 | 0 | | 2.4–3.4 | 11 d | | 15.0 | 85 | 0.55 | |
| | | | | 75 | | 2.1–3.4 | 6–11 d | | 23.6*+ | 78 | 0.50 | |
| | | | 4.0 | 0 | | 2.8–3.7 | 1–7 d | | 19.2+ | 70 | 5.22+ | |
| | | | | 75 | | 2.3–3.7 | 4–11 d | | 29.2+ | 81 | 1.10* | |
| | | 15 | 6.5 | 0 | | 0.65 | 8 d | 37 | | | | 2 |
| | | | 5.2 | | | 0.68 | | 36 | | | | |
| | | | | 27 | | 0.55 | | 42* | | | | |
| | | | 6.5 | 0 | | 0.65 | 21 d | 35 | | | | |
| | | | 5.2 | | | 0.68 | | 33 | | | | |
| | | | | 27 | | 0.54 | | 37 | | | | |
| | | | | 118 | | 0.60 | 40 hr[2] | 47 | | | | |
| | | | | | | | | 42* | | | | |
| | | | | | | | | 45* | | | | |
| *Oncorhynchus mykiss* | 1-yr | 10 | 5.2 | 0 | | | 96 hr | 38 | | | | 3 |
| | | | | 137 | 92 | | | 45* | | | | |
| | | | | 203 | 175 | | | 54* | | | | |
| | | | 5.4 | 0 | | | | 38 | | | | |

| | | | | | | | | | | | | |
|---|---|---|---|---|---|---|---|---|---|---|---|---|
| | | | | 142 | 123 | | 96 hr | 41 | | | | 3 |
| | | | | 0 | | | | 50 | | | | |
| | | | | 490 | 193 | | | 61* | | | | |
| | | | 5.6 | 0 | | | | 51 | | | | |
| | | | | 173 | 63 | | | 56 | | | | |
| | | | | 0 | | | | 46 | | | | |
| | | | | 530 | 341 | | | 61* | | | | |
| *Oncorhynchus mykiss* | adult | 14 | 4.8 | 0 | | 1.00 | 0 hr | 27.2 | | | 0 | 4 |
| | | | | | | | 24 hr | 15.8* | | | 0 | |
| | | | | | | | 72 hr | 13.6* | | | 0 | |
| | | | | 100 | | | 0 hr | 23.7 | | | 0 | |
| | | | | | | | 24 hr | 25.8 | | | 4.2* | |
| | | 13 | 5.0 | 0 | | 132 | 6 hr | 24.2 | 7.8 | 108 | | 5 |
| | | | | | | | 4 d | 16.8 | 7.8 | 108 | | |
| | | | | 2000 | | | 6 hr | 20.5 | 7.7 | 115 | | |
| | | | | | | | 4 d | 32.0 | 7.45 | 42 | | |
| | | 15 | 6.5 | 5 | | 0.9 | 0 hr | | 7.78 | 106 | 0.52 | 6 |
| | | | 5.2 | | | | ≤66 hr | | 7.67 | 105 | 0.43 | |
| | | | | 105 | | | | | 7.59 | 22* | 4.42* | |
| | | | | 5 | | 8.2 | | | 7.81 | 120 | 0.95 | |
| | | | | 105 | | | | | 7.72* | 21* | 4.63* | |
| | | | 4.8 | 5 | | 0.9 | | | 7.66 | 117 | 0.52 | |
| | | | | 105 | | | | | 7.38* | 39* | 2.81 | |
| | | | | 5 | | 8.2 | | | 7.62 | 133 | 1.26 | |
| | | | | 105 | | | | | 7.63 | 102 | 0.69 | |
| | | | 4.4 | 5 | | 0.9 | | | 7.77 | 111 | 0.83 | |
| | | | | 105 | | | | | 7.45* | 81* | 1.28 | |
| | | | | 5 | | 8.2 | | | 7.56 | 105 | 1.78 | |
| | | | | 105 | | | | | 7.43* | 111 | 1.28 | |

(continued)

Table 9. (*Continued*)

| Species | Life stage | Experimental conditions: Temp (°C) | pH | Total Al | Available Al | $Ca^{2+}$ | Exposure time | Blood chemistry: Hct (%) | pH | $PO_2$ | Lactate | Reference |
|---|---|---|---|---|---|---|---|---|---|---|---|---|
| | | 16 | 5.0 | 0 | | 40 | 0 hr | 21 | 7.70 | 132 | | 7 |
| | | | | 60 | | | | 21 | 7.70 | 132 | | |
| | | | | 0 | | | 28 hr | 18 | 7.60 | 136 | | |
| | | | | 60 | | | | 24* | 7.41* | 102* | | |
| | | | | 0 | | | 52 hr | 16.5 | 7.60 | 127 | | |
| | | | | 60 | | | | 25* | 7.35* | 92* | | |
| | | 9 | | 0 | | | 0 d | 32 | | | | 8 |
| | | | | 200 | | | 3 d | 53* | | | | |
| | | | | 0 | | | 0 d | 22 | | | | |
| | | | | 60 | | | 1 d | 30* | | | | |
| | | | | 0 | | 3.50 | 0 d | 24[a] | | | | 9 |
| | | | 4.7 | 180 | | | 2 d | 51[b] | | | | |
| | | 17 | | 0 | | 0.68 | 3.5 hr | 41 | | | | 10 |
| | | | | 351 | | | | 45 | | | | |
| | | | | | | 3.51 | | 42 | | | | |
| *Salmo salar* | parr | 4–6 | 5.5 | 140[3] | | | 27 d | 33 | | | | 11 |
| | | | | 240 | | | 29 d | 43* | | | | |
| | | | 5.3 | 155 | | | 9 d | 43* | | | | |
| | | | | 160 | | | 11 d | 42* | | | | |
| | | | | 140 | | | 27 d | 43* | | | | |
| | | | | 240 | | | 29 d | 44* | | | | |
| *Salmo trutta* | juvenile | 10 | 4.5 | 0 | | | 30 hr | 41 | | | | 12 |
| | | | | 324 | | | | 48* | | | | |

| | | | | | | | | | | | | |
|---|---|---|---|---|---|---|---|---|---|---|---|---|
| | | | | 0 | | | | 43 | | | | |
| | | | | 324 | | | | 49 | | | | |
| *Salmo trutta* | adult | 8 | 5.1 | 360 | | | 50 hr | 33 | | | | 11 |
| | | | | | | | | 48* | | | | |
| | | | | 13 | | | 142 hr | 48* | | | | |
| *Salvelinus fontinalis* | juvenile | 10 | 5.2 | 0 | | 0.56 | 24 d | 34.9 | | | 5.9 | 13 |
| | | | | 150 | | | 4 d | 46.0* | | | 10.8* | |
| | | | | 75 | | | 21 d | 42.0* | | | 5.9 | |
| *Salvelinus fontinalis* | adult | 11 | 5.2 | 0 | | 0.50 | 10 wk | | 7.94 | 52 | 2.65 | 14 |
| | | | | 333 | | | 8 hr | | 7.79* | 54 | 2.80 | |
| | | | | | | | 24 hr | | 7.58* | 46 | 5.23 | |
| | | | | 75 | | | 10 wk | | 7.70 | 54 | 2.60 | |
| | | | | 333 | | | 8 hr | | 7.70 | 58 | 4.30 | |
| | | | | | | | 24 hr | | 7.72 | 51 | 2.60 | |
| | | 10 | 4.8 | 0 | | 0.50 | 42 hr | | 0.1* | +9 | +0.31[12] | 15 |
| | | | | 333 | 233 | | | | −0.09 | −2 | +1.29* | |
| | | | | | | 8.0 | | | −0.18* | −50* | +2.19* | |
| | | | 4.4 | 0 | | 0.50 | | | −0.11* | +11 | +0.75* | |
| | | | | 333 | 233 | 8.0 | | | −0.20* | −9 | +0.92 | |
| | | 11 | 6.5 | 4 | 4 | 0.50 | 10 wk | 43 | | | | 16 |
| | | | 4.8 | 3 | 3 | | 48 hr | 36 | | | | |
| | | | | 333 | | | | 62 | | | | |
| | | | 5.2 | 3 | 3 | | 10 wk | 46 | | | | |
| | | | 4.8 | | | | 48 hr | 44 | | | | |
| | | | 5.2 | 77 | 89 | | 10 wk | 31 | | | | 16 |
| | | | 4.8 | 333 | | | 48 hr | 35 | | | | |
| | | | 5.2 | 136 | 134 | | 10 wk | 35 | | | | |
| | | | 4.8 | 333 | | | 48 hr | 41*[4] | | | | |
| | | | 6.5 | 4 | 3 | 8.00 | 10 wk | 38 | | | | |

(continued)

Table 9. (*Continued*)

| Species | Life stage | Experimental conditions: Temp (°C) | pH | Total Al | Available Al | $Ca^{2+}$ | Exposure time | Blood chemistry: Hct (%) | pH | $PO_2$ | Lactate | Reference |
|---|---|---|---|---|---|---|---|---|---|---|---|---|
| | | | 4.8 | 0 | | | 48 hr | 42 | | | | |
| | | | | 333 | | | | 54* | | | | |
| | | | 5.2 | 3 | 3 | | 10 wk | 40 | | | | |
| | | | 4.8 | 333 | | | 48 hr | 40 | | | | |
| | | | 5.2 | 81 | 73 | | 10 wk | 40 | | | | |
| | | | 4.8 | 333 | | | 48 hr | 36 | | | | |
| | | | 5.2 | 136 | 134 | | 10 wk | 41 | | | | |
| | | | 4.8 | 333 | | | 48 hr | 40 | | | | |
| | | 10 | 6.5 | 0 | | 0.50 | 10 wk | | 7.96 | 65.6 | 1.07 | 17 |
| | | | 4.8 | 333 | | | 66 hr | | 7.72[5] | | | |
| | | | 6.5 | 0 | | 8.00 | 10 wk | | 7.77 | 76.4 | 0.78 | |
| | | | 4.8 | 333 | | | 66 hr | | 7.57+ | | | |
| | | | 5.2 | 0 | | 0.05 | 10 wk | | 7.91 | 62.8 | 0.78 | |
| | | | 4.8 | 333 | | | 66 hr | | 7.74+ | | | |
| | | | 5.2 | 150 | | | 10 wk | | 7.80 | 68.7 | 0.59* | |
| | | | 4.8 | 333 | | 0.05 | 66 hr | | 7.76+ | | | |

1. Results that are significantly different from results observed at the same pH and 0 $\mu g \cdot L^{-1}$ Al are indicated by (*); those that are significantly different from a circumneutral control (usually pH 6.5 or 7.0) and 0 $\mu g \cdot L^{-1}$ Al are indicated by (+).

2. After 21-d exposure to above conditions, fish from all treatments were challenged with a 40 hr exposure to 118 $\mu g \cdot L^{-1}$ of Al (Reid et al., 1991) .

3. Represents midway points on the following ranges: 110–170, 230–250, 130–180, 140–180, 110–170, and 230–250 $\mu g \cdot L^{-1}$ Al, respectively.

4. Result significantly different from the result measured after 10-wk exposure shown immediately above.

5. Result significantly different from the result measured after 10-wk exposure shown immediately above.

References: (1) Neville 1985; (2) Reid et al. 1991; (3) Dietrich and Schlatter 1989b; (4) Goss and Wood 1988; (5) Malte 1986; (6) Playle et al. 1989; (7) Witters et al. 1991; (8) Witters et al. 1990a; (9) Witters et al. 1990b; (10) Witters 1986; (11) Fivelstad and Leivestad 1984; (12) Reader et al. 1991; (13) McDonald et al. 1991; (14) Walker et al. 1991; (15) Wood et al. 1988a; (16) Wood et al. 1988b; (17) Wood et al. 1988c.

Table 10. Aluminum concentrations ($\mu g \cdot g^{-1}$) in various tissues of fish experimentally exposed to aqueous Al. Concentrations are expressed on a wet-weight basis unless otherwise noted. Species name, life stage, and exposure condition cells with missing blanks have the same value as the cell above within the same study.

| | Exposure conditions | | | | | Aluminum concentration | | | | | | |
|---|---|---|---|---|---|---|---|---|---|---|---|---|
| Species | Life stage | Temp (°C) | pH | Al Level | Exposure time | Whole body | Gills | Gut | Kidney | Liver | Muscle | Reference |
| *Oncorhynchus mykiss* | juvenile | 0.5 | 5.5 | 100 | 7 d | | 1.8[1] | | | | | 1 |
| | | | | 300 | | | 8.7 | | | | | |
| | | | | 600 | | | 12.5 | | | | | |
| | | 10 | 6.5 | 75 | 11 d | | 275 | | | | | 2 |
| | | | 6.1 | | | | 490 | | | | | |
| | | | 5.0 | | | | 121 | | | | | |
| | | | 4.0 | | | | 30 | | | | | |
| | | | 6.0 | —[2] | 3 hr | 32.5 | | | | | | 3 |
| | | | 7.65 | | 96 hr | 3.9 | | | | | | |
| | | | 9.00 | | | 6.0 | | | | | | |
| | | | 10.00 | | 3 hr | 22.6 | | | | | | |
| *Oncorhynchus mykiss* | adult | | 4.5–6.0 | | 28 hr | | 60 | | | | | 4 |
| | | | 4.1 | 428 | 24 hr | | 105 | | | | | 5 |
| | | | 5.4 | 34.5 | | | 7 | | | | | |
| | | | 4.2 | 37 | | | 28 | | | | | |
| | | 15 | 5.2 | 105 | 66 hr | | 20 | | | | | 6 |
| | | | 4.8 | | | | 18 | | | | | |
| | | | 4.4 | | | | 4 | | | | | |
| | | 8–10 | 4.7 | 0 | 10 d | 0.9 | | | | | | 7 |
| | | | | 179 | | 75.0 | | | | | | |

*(continued)*

Table 10. (*Continued*)

| Species | Exposure conditions | | | | | Aluminum concentration | | | | | | Reference |
|---|---|---|---|---|---|---|---|---|---|---|---|---|
| | Life stage | Temp (°C) | pH | Al Level | Exposure time | Whole body | Gills | Gut | Kidney | Liver | Muscle | |
| | | 8 | 5.0 | 200 | 3 hr | | 38 | | | | | 8 |
| | | | | | 3 d | | 81.4 | | | | | |
| *Salmo salar* | eggs | 0.5 | 5.5 | 300 | 14 d | 49 | | | | | | 1 |
| | | | | 100 | | 31 | | | | | | |
| | | | | 300 | | 15.5 | | | | | | |
| | | | | 100 | | 15.0 | | | | | | |
| *Salmo salar* | alevin | 10 | 5.0 | 100 | 18 d | 69[3] | | | | | | 9 |
| | | | | | | 40[4] | | | | | | |
| *Salmo trutta* | alevin | 12 | 5.0 | 230 | 5 d | 51 | | | | | | 10 |
| *Salmo trutta* | juvenile | | 6.4 | 99 | 17 d | | 61[1] | | | | | 11 |
| | | | 4.9 | 450 | | | 1582 | | | | | |
| | | | 4.8 | 216 | | | 673 | | | | | |
| | | | 5.3 | 333 | | | 2197 | | | | | |
| | | | 5.3 | 126 | | | 406 | | | | | |
| | | | 4.6 | 216 | | | 1275 | | | | | |
| | | | 4.4 | 513 | | | 3505 | | | | | |
| *Salmo trutta* | 1 yr | 11 | 5.08 | 75 | 2 d | | 124.7 | | | | | 12 |
| | | 8.5 | 5.7 | 151 | 5 d | | 367.3 | | | | | |
| | 2 yr | 11 | 5.08 | 75 | 2 d | | 117.5 | | | | | |
| | | 8.5 | 5.7 | 151 | 5 d | | 210.2 | | | | | |

| | | | | | | | | |
|---|---|---|---|---|---|---|---|---|
| | 1 yr | 2.5 | 5.5 | 0 | 42 d | | 10 | 13 |
| | | | | 50 | | | 65 | |
| | | | | 200 | | | 140 | |
| | | | | 500 | | | 95 | |
| | | | 7.0 | 0 | | | 10 | |
| | | | | 50 | | | 13 | |
| | | | 7.0 | 200 | | | 21 | 13 |
| | | | | 500 | | | 19 | |
| | | 15 | 5.5 | 0 | | | 0 | |
| | | | | 50 | | | 36 | |
| | | | | 200 | | | 67 | |
| | | | | 500 | | | 97 | |
| | | | 7.7 | 0 | | | 0 | |
| | | | | 50 | | | 11 | |
| | | | | 200 | | | 21 | |
| | | | | 500 | | | 43 | |
| *Salvelinus fontinalis* | alevin | 12 | 7.2 | 415 | 3 d | 12.5[a] | | 14 |
| | | | | 305 | 7 d | 8.3[a] | | |
| | | | | 267 | 56 d | 3.8[a] | | |
| | | | 6.1 | 324 | 3 d | 18.5[a] | | |
| | | | | 267 | 7 d | 40.7[b] | | |
| | | | | 217 | 56 d | 16.6[a] | | |
| | | | 5.3 | 252 | 3 d | 58.4[a] | | |
| | | | | 239 | 7 d | 46.4[ab] | | |
| | | | | 207 | 56 d | 37.3[b] | | |
| | | 11 | 5.2[6] | 333 | 21 d | 189 | | 15 |
| | | | 4.8[6] | | | 149 | | |

*(continued)*

Table 10. (*Continued*)

| | Exposure conditions | | | | | Aluminum concentration | | | | | | |
|---|---|---|---|---|---|---|---|---|---|---|---|---|
| Species | Life stage | Temp (°C) | pH | Al Level | Exposure time | Whole body | Gills | Gut | Kidney | Liver | Muscle | Reference |
| *Salvelinus fontinalis* | juvenile | 10 | 5.2 | 150 | 24 hr | | 251 | | | | | 16 |
| | | | | 150[7] | 24 d | | 40 | | | | | |
| | | 11 | | 111[8] | 91 d | 127 | | | | | | 15 |
| | | | | 333[8] | | 254 | | | | | | 17 |
| | | | | | | 230 | | | | | | 15 |
| | | | 4.8 | | | 100 | | | | | | |

1. Dry weight concentrations.
2. Fish exposed to alum sludge bearing 108,000 $\mu g \cdot kg^{-1}$ of Al.
3. An average of 42% of Al adsorbed on outside of fish. Average weight of fish, 0.02 g.
4. An average of 25% of Al adsorbed on outside of fish. Average weight of fish, 0.04 g.
5. Letters indicate results that are the same or different among various exposure times within the same pH and Al concentration.
6. Calcium concentrations, 0.5 $mg \cdot L^{-1}$.
7. Fish exposed for 3 d to 150 $\mu g \cdot L^{-1}$ followed by 21-d exposure to 75 $\mu g \cdot L^{-1}$.
8. Calcium concentrations, 1.0 $mg \cdot L^{-1}$.

References: (1) Brouard et al. 1989; (2) Neville and Campbell 1988; (3) Ramamoorthy 1988; (4) Harvey and Whelpdale 1986; (5) Harvey and McArdle 1986; (6) Playle et al. 1989; (7) Witters et al. 1990b; (8) Witters et al. 1987; (9) Parent et al. 1988; (10) Segner et al. 1988; (11) Stoner et al. 1984; (12) Dietrich and Schlatter 1989a; (13) Karlsson-Norrgren et al. 1986b; (14) Cleveland et al. 1991a; (15) Wood et al. 1990b; (16) McDonald et al. 1991; (17) Wood et al. 1990a.

fish captured in water of pH 9 and $\leq 420\ \mu g \cdot L^{-1}$ Al (Table 11) (Buergel and Soltero 1983). Gill tissue Al levels in these fish ($5.9–13.6\ \mu g \cdot g^{-1}$) were generally one to two orders of magnitude higher than in gut tissue ($0.05–1.82\ \mu g \cdot g^{-1}$), the tissue with the second highest concentration.

**Atlantic Salmon** (*Salmo salar*) **and Brown Trout** (*Salmo trutta*). The Atlantic salmon is an anadromous species that spawns in coastal drainages of northeastern North America and western Europe (Lee et al. 1980). Thus, most studies have examined exposure to elevated Al during the early life stages. The brown trout is an anadromous and freshwater species native to Europe and western Asia but has been stocked widely in North America. Research has included all life stages.

Fivelstad and Leivestad (1984) reported $LT_{50}$s of 150–62 hr among Atlantic salmon alevins exposed to pH 5.1 and $100–230\ \mu g \cdot L^{-1}$ Al, respectively. Among parr exposed to pH of 4.9–5.3, $LT_{50}$s decreased from 89 hr to 26 hr as total Al increased from 140 to $420\ \mu g \cdot L^{-1}$ (Table 7). Parr exposed to pH of 5.3–5.5 and $140–240\ \mu g \cdot L^{-1}$ Al generally had higher hematocrit counts than those not exposed (Table 9). All smolts died by 48 hr of exposure in pH $\leq 5.6$ and $268\ \mu g \cdot L^{-1}$ Al, and only 30% survived at pH 6.05 and $268\ \mu g \cdot L^{-1}$ (Skogheim et al. 1986). All adults died by 18 d exposure to pH 5.0 and $163–277\ \mu g \cdot L^{-1}$ Al. The $LT_{50}$s for the adults exposed to these Al levels ranged from 109 hr to 32 hr, respectively (Skogheim and Rosseland 1986).

Whole-body Al has been measured only in Atlantic salmon eggs under conditions free of dissolved organic carbon (DOC) and F (Table 10). After 14 d exposure to $100–300\ \mu g \cdot L^{-1}$ Al, fertilized eggs accumulated $5–15.5\ \mu g \cdot g^{-1}$ of Al, whereas embryonated eggs accumulated $29–40\ \mu g \cdot g^{-1}$ (Brouard et al. 1989).

Mortality studies suggest that brown trout alevins are sensitive to Al levels ranging from about 140 to $250\ \mu g \cdot L^{-1}$. However, Ca may affect this sensitivity (Table 7). Survival was severely reduced at $250\ \mu g \cdot L^{-1}$ Al and pH ranging from 5.4 to 4.5 with $0.5\ mg \cdot L^{-1}$ Ca. Under similar conditions but $1.0\ mg \cdot L^{-1}$ Ca, Al reduced $H^+$ toxicity. At $250\ \mu g \cdot L^{-1}$ Al, survival increased progressively as pH dropped from 5.4 to 4.5 (Brown 1983). However, in another study (Sayer et al. 1991b), dissolved Al seemed to reduce survival at low ($<1\ mg \cdot L^{-1}$) and high ($>7\ mg \cdot L^{-1}$) Ca levels.

Most of the studies on juvenile brown trout have focused on blood parameters. Reader et al. (1991) demonstrated that a 30-hr exposure to pH 4.5 and $314\ \mu g \cdot L^{-1}$ Al reduced plasma electrolyte concentrations for up to 120 hr after fish were returned to water of pH 5.6 (Table 8). The hematocrit count 24 hr after exposure was significantly higher than controls, but at 120 hr after exposure it had returned to control levels (Table 9). Battram (1988) showed that $175\ \mu g \cdot L^{-1}$ Al caused a net loss in Cl at pH 7.0 and appeared to exacerbate losses at pH 5.5.

Mortality studies suggest that adult brown trout may be nearly as sensi-

Table 11. Aluminum concentrations ($\mu g \cdot g^{-1}$) in various tissues of wild-caught fish and associated ambient Al ($\mu g \cdot L^{-1}$) and pH values.

| Species | Life stage | pH range | Ambient Al | Whole body | Bone | Gill | Gut | Kidney | Liver | Muscle | Other | References |
|---|---|---|---|---|---|---|---|---|---|---|---|---|
| *Dorosoma cepedianum* | adult | 7.0–8.0 | ≤100 | | | 229 | | 551 | 1138 | 39.3 | 1548[1] | 1 |
| *Oncorhynchus mykiss* | juvenile | 9.0 | ≤420 | | | 6.3[2] | 0.70 | 0.39 | 0.59 | 0.18 | 0.19[3] | 2 |
| | | | | | | 13.5 | 0.79 | 0.36 | 0.16 | 0.01 | 0.27 | |
| | | | | | | 9.03 | 1.82 | 0.31 | 0.58 | 0.00 | 1.78 | |
| *Oncorhynchus mykiss* | 1+ yr | | | | | 5.88 | 0.70 | 1.03 | 1.41 | 0.05 | 0.28 | |
| | | | | | | 12.4 | 0.05 | 0.12 | 0.32 | 0.01 | 0.55 | |
| | | | | | | 13.6 | 1.91 | 0.19 | 0.20 | 0.01 | 0.05 | |
| *Oncorhynchus mykiss* | 2+ yr | | | | | 6.08 | 0.67 | 2.51 | 3.99 | 0.14 | 0.38 | |
| | | | | | | 7.25 | 0.50 | 0.04 | 0.08 | 0.01 | 1.02 | |
| *Salmo trutta* | adult | 5.37 | | 168 | | 126.9 | | 12.9 | 30.5 | | | 3 |
| | | 6.9 | 35 | | 1.9 | 1.9 | 2.6 | 1.7 | 0.8 | 0.6 | | 4 |
| | | 6.3 | 288 | | 6.5 | 17.6 | 7.3 | 9.9 | 1.9 | 1.8 | | |
| | | 6.9 | 208 | | 6.0 | 19.4 | 5.0 | 7.6 | 4.6 | 2.6 | | |
| | | 5.8 | 216 | | 4.4 | 89.3 | 4.4 | 1.5 | 1.3 | 0.8 | | |
| *Salvelinus fontinalis* | 2–4 yr | 5.2–5.5 | 250[4] | | | 11.4 | | | | | | 5 |
| | | | <50 | | | 19.0 | | | | | | |
| | adult | 5.18 | 91 | 7.74 | | | | | | | | 6 |
| | | 5.43 | 286 | 8.85 | | | | | | | | |
| | | 6.63 | 245 | 2.03 | | | | | | | | 7 |
| | | 5.80 | 45 | 6.75 | | | | | | | | |
| | | 4.95 | 169 | 12.70 | | | | | | | | |
| | | 5.08 | 287 | 4.43 | | | | | | | | |

| | | | | | | | | | | | | |
|---|---|---|---|---|---|---|---|---|---|---|---|---|
| *Catostomus commersoni* | adult | 6.2–6.3 | 15 | | 15 | | | | | | | |
| | | | 50 | | 8 | | | | | | | |
| | | 4.8–5.8 | 120 | | 22 | | | | | | | |
| | | | 50 | | 12 | | | | | | | |
| | | | 105 | | 22 | | | | | | | |
| | | 5.0 | 90 | | 17 | | | | | | | |
| | | 4.8 | 180 | | 15 | | | | | | | |
| | | 7.02 | 28 | 2.45[2] | | | | | | | | 8 |
| | | 5.83 | 66 | 6.07 | | | | | | | | |
| | | 5.61 | 51 | 4.01 | | | | | | | | |
| | | 5.43 | 163 | 3.18 | | | | | | | | |
| *Ictalurus punctatus* | adult | 7.0–8.0 | 0 | | | 284 | | 1739 | 1809 | 60.8 | 1478[3] | 1 |
| | | | 100 | | | 108 | | 489 | 726 | | | |
| *Micropterus dolomieui* | n/r[5] | 6.3 | ≤940 | 3.9 | | 58 | 2.5 | <1.0 | 1.5 | | | 9 |
| *Micropterus salmoides* | adult | 7.0–8.0 | ≤100 | | | 148 | | 1048 | 915 | 72.1 | 1509[1] | 1 |
| *Perca flavescens* | adult | 4.7 | 35 | | | 6.5 | | | | | | 10 |
| | | 5.1 | 27 | 4.1 | | | 67.0 | | | 1.0 | 3.5[6] | |
| | | 6.1 | 16 | 3.6 | | 11.0 | 69.0 | | | 0.85 | 7.0[6] | |

1. Brain tissue.
2. Values reported on a dry-weight basis.
3. Heart tissue.
4. Median for range of 200–300 $\mu g \cdot L^{-1}$ Al.
5. Not reported (n/r).
6. Skin tissue.

References: (1) Berg and Burns 1985; (2) Buergel and Soltero 1983; (3) Dietrich and Schlatter 1989a; (4) Karlsson-Norrgren et al. 1986a; (5) Chevalier et al. 1985; (6) Haines et al. 1987; (7) Bendell-Young and Harvey 1986; (8) Hamilton and Haines 1989; (9) Brumbaugh and Kane 1985; (10) King et al. 1992.

tive as rainbow trout. The $LT_{50}$s of adult brown trout exposed to pH 5.2 and 105 $\mu g \cdot L^{-1}$ Al ranged from 16 hr in water with low dissolved NaCl to 85 hr in water with higher levels of NaCl (Dietrich et al. 1989). Marked reductions in survival occurred among adults exposed for 6 wk to pH 5.2 and either 54 or 81 $\mu g \cdot L^{-1}$ Al (Sadler and Lynam 1988).

Short-term protective effects of Al against $H^+$ toxicity have been shown in some studies. Levels of 250 $\mu g \cdot L^{-1}$ Al more than doubled survival time among adults exposed to pH 4.0 and 3.5; however, the $LT_{50}$s of adult fish exposed to pH 3.5 were all less than 24 hr, regardless of dissolved Al levels (Brown 1981).

Both the detrimental and protective aspects of dissolved Al are reflected in patterns of net ion fluxes of adult brown trout exposed for short periods to elevated Al (Table 12). Adults exposed for 2 hr to 216 $\mu g \cdot L^{-1}$ at pH 7 and 5.4 experienced greater net Na losses than those exposed to similar pH without Al. However, after several hours of exposure, the fish seemed to develop resistance to the adverse effects of elevated Al. After 8 hr, net ion fluxes were nearly 0 in all treatments. The protective effects of Al were observed in fish exposed to 216 $\mu g \cdot L^{-1}$ and pH 4. After 2-hr and 8-hr exposures, net Na losses were lower in fish when Al was present than when it was not. Small increases in dissolved Ca generally reduced Na losses. Net Na losses were lower in fish exposed to 2.0 $mg \cdot L^{-1}$ Ca than in those exposed to 0.4 $mg \cdot L^{-1}$ Ca, regardless of Al (Dalziel et al. 1986).

Aluminum concentrations in brown trout gills were highest at the lowest pH, but in other tissue it declined with pH (Table 11) (Karlsson-Norrgren et al. 1986a). Gill tissue residues may also increase with water temperatures (Table 10) (Karlsson-Norrgren et al. 1986b), perhaps because metabolic rates are higher in warmer water.

**Brook Trout** (*Salvelinus fontinalis*). Brook trout are native to clear cold-water streams and lakes of eastern North America (Lee et al. 1980). Much of their native and introduced habitat lies within some of the areas in North America that are most vulnerable to acid deposition (Haines 1981). Because of their geographic distribution and habitat requirements, brook trout have been the most frequently studied fish species with regard to acid deposition and elevated dissolved Al.

Results on brook trout eggs exposed to various pH and Al levels leave a mixed impression of the role that the metal plays in egg survival. Two studies (Baker and Schofield 1982; Ingersoll et al. 1990a) indicate that 200–500 $\mu g \cdot L^{-1}$ Al (pH 4.2–4.9) improved survival in eyed eggs exposed (Table 7). Other studies indicate that elevated Al increased the frequency of incomplete hatching of brook trout eggs at pH greater than 5.6. Within the pH range of 5.6–6.6 and Al of 4–350 $\mu g \cdot L^{-1}$, 92%–96% of eyed eggs hatched regardless of Al level. However, the percentage of incompletely hatched eggs progressively increased with increasing levels (Cleveland et al. 1989). The frequency of incomplete hatching also increased in 300 $\mu g \cdot L^{-1}$ Al at pH 7.2 and 5.5. At pH 4.5, there was more incomplete hatching without than with Al exposure (Cleveland et al. 1986).

Alevin and juvenile survival and growth have been studied using various Ca, pH, and Al levels and exposure times (Table 7). Alevin and juvenile survival decreased from exposure to pH 4.4 and 1187 $\mu g \cdot L^{-1}$ Al, but not from exposures to pH 4.4–5.2, $\leq 350$ $\mu g \cdot L^{-1}$ Al, and 1.48 $mg \cdot L^{-1}$ Ca. At pH 4.0, all alevins and juveniles died (Ingersoll et al. 1990a). Alevin and juvenile survival decreased as Al levels increased at pH $\leq 4.6$ (Baker and Schofield 1982). Cleveland et al. (1986) also showed that Al tends to further decrease survival at pH 4.5–5.5 by exacerbating electrolyte losses (Neville and Campbell 1988).

Alevin survival in the presence of 200–500 $\mu g \cdot L^{-1}$ Al was lower at pH near 5.5 than at pH 7 or 4.9 (Baker and Schofield 1982), with a toxic threshold level between 140 and 350 $\mu g \cdot L^{-1}$ at pH 5.5–6.5. Mortality rates did not change significantly in alevins exposed to 8–142 $\mu g \cdot L^{-1}$ at pH 5.6 nor in alevins exposed to 4–169 $\mu g \cdot L^{-1}$ at pH 6.5. However, mortalities increased significantly at 292 $\mu g \cdot L^{-1}$ and pH 5.6, and at 350 $\mu g \cdot L^{-1}$ with pH 6.6 (Cleveland et al. 1989). These Al levels were more than sixfold greater than the saturation level (Skogheim et al. 1987).

Alevin survival decreases as length of exposure to low pH and elevated Al increases. Percentage survival rates in alevins exposed to pH 5.6 and 296 $\mu g \cdot L^{-1}$ Al were 97.7% at 15 d, 73% at 30 d, 42.3% at 45 d, and 29.7% at 60 d. Survival of alevins exposed to pH 6.6 and 350 $\mu g \cdot L^{-1}$ was 96.2% at 15 d, 91.5% at 30 d, 62.5% at 45 d, and 51.5% at 60 d (Table 7) (Cleveland et al. 1989).

Brook trout alevin growth rates appear to be more sensitive than survival to reduced pH and elevated Al. In three studies, growth declined with pH, regardless of Al levels (Cleveland et al. 1986, 1989, 1991a). Baker and Schofield (1982), however, showed that both increased Al and acidity reduce alevin and juvenile growth rates (Table 7).

A single study on electrolyte concentrations in brook trout alevins exposed to Al produced contradictory results. Alevins were exposed to 300–453 $\mu g \cdot L^{-1}$ Al and pH 4.0–7.2 for 160 hr. Exposures were divided into four 40-hr pulses separated by 5 d at no Al and pH 7.0. Whole-body Na and K concentrations in all treatments with Al were higher than in those at similar pH without Al. If the alevins had responded to the treatments according to the Neville and Campbell (1988) model, electrolyte concentrations during Al exposure should have been identical or lower than when Al was absent (Table 8) (Cleveland et al. 1991b). Juveniles appeared to follow the Neville and Campbell (1988) model in that their Na levels declined upon Al exposure (Table 8) (Gagen and Sharpe 1987; McDonald et al. 1991). Hematocrits also were higher in Al-exposed than in unexposed juveniles (Table 9).

Low Ca levels can increase brook trout survival from fertilization through the early juvenile stages. Although survival was not reduced at Al $\leq 350$ $\mu g \cdot L^{-1}$, pH $\geq 5.2$, and 0.37–8.0 $mg \cdot L^{-1}$ Ca, it dropped substantially

Table 12. Influence of dissolved organic carbon (DOC) ($mg \cdot L^{-1}$) on percentage survival of fish experimentally exposed to different levels of Al ($\mu g \cdot L^{-1}$) and Ca ($mg \cdot L^{-1}$).

| | | Experimental conditions | | | | | | | | | |
|---|---|---|---|---|---|---|---|---|---|---|---|
| Species | Life stage | Temp (°C) | pH | Total Al | Available Al | $Ca^{2+}$ | DOC | Organic compound | Expo-sure time | Survival (%) | Refer-ence |
| *Oncorhynchus mykiss* | adult | 9 | 4.7 | 179 | 175 | 3.5 | 0 | humic materials | 3 d | 0 | 1 |
| | | | | 184 | 36 | | 9.4 | | | 100 | |
| *Salmo salar* | alevin | 15 | 5.0 | 23 | 5 | 2.0 | <0.5 | silicic acid | 96 hr | 100 | 2 |
| | | | | 181 | 135 | | <0.5 | | 60 hr | 0 | |
| | | | | 194 | 162 | | 2.52 | | 96 hr | 100 | |
| | | 6 | 4.8 | 0 | | 0.4 | 0 | citric acid | 43 d | 15 | 3 |
| | | | | | | | 0.25 | | | 25 | |
| | | | | | | | 2.50 | | | 0 | |
| | | | | 181 | | | 0 | | | 0 | |
| | | | | | | | 0.25 | | | 0 | |
| | | | | | | | 2.50 | | | 85 | |
| | | | | 1809 | | | 0 | | | 0 | |
| | | | | | | | 0.25 | | | 0 | |
| | | | | | | | 2.50 | | | 0 | |
| *Salmo salar* | parr | 11 | 4.9 | 383 | 113 | 3.5 | 0 | citric acid | 24 hr | 58.8 | 4 |
| | | | | 232 | 0 | | 0.5–0.7 | | | 81.8 | |
| | | | | 383 | 113 | | 0 | | | 0 | |
| | | | | 232 | 0 | 3.5 | 0.5–0.7 | | | 100 | |
| *Salmo trutta* | juvenile | | | 383 | 113 | 3.5 | 0 | | | 50.0 | |
| | | | | 232 | 0 | | 0.5–0.7 | | | 100 | |
| | | | | 383 | 113 | | 0 | | | 35.7 | |
| | | | | 232 | 0 | | 0.5–0.7 | | | 92.7 | |

| | | | | | | | | | | | |
|---|---|---|---|---|---|---|---|---|---|---|---|
| *Salvalinus fontinalis* | alevin | 9–17 | 5.2 | 420 | | 2–4 | 0 | citric acid | 14 d | 28 | 5 |
| | | | | 500 | | | 11.7 | | | 96 | |
| | | | 4.4 | 480 | | | 0 | | | 42 | |
| | | | | 500 | | | 11.7 | | | 82 | |
| *Salvalinus fontinalis* | juvenile | 12 | 5.6 | 1174 | 649 | 2.2 | 0.6–5.0 | humic material | 21 d | 0 | 6 |
| | | | | 134 | 31 | | 0.6 | | | 10.6 | |
| | | | | 11 | 1 | | | | | 74.0 | |
| | | | | 339 | 60 | | 3.1 | | | 86.4 | |
| | | | | 1174 | 160 | | 9.2 | | | 38.0 | |
| | | | 5.2 | 1176 | | | 0.6–8.3 | | | 0 | |
| | | | | 335 | 177 | | 0.6 | | | | |
| | | | | 121 | 10 | | | | | 50.6 | |
| | | | | 335 | 59 | | 3.5 | | | 67.2 | |
| | | | 4.8 | 1276 | | | 0.6–7.5 | | | 0 | |
| | | | | 339 | 201 | | 0.6 | | | 13.6 | |
| | | | | 135 | 63 | | | | | 64.2 | |
| *Salvalinus fontinalis* | | | | 339 | 17 | 2.2 | 7.5 | | | 58.5 | |
| | | | 4.4 | 11 | 1 | | 0.6 | | | 42.6 | |
| | | | | | | | 3.1 | | | 17.7 | |
| | | | | | | | 5.5 | | | 10.9 | |
| | | | | | | | 7.8 | | | 30.5 | |

References: (1) Witters et al. 1990b; (2) Birchall et al. 1989; (3) Peterson et al. 1989; (4) McCahon & Pascoe 1989; (5) Driscoll et al. 1980; (6) Parkhurst et al. 1990.

when juveniles were exposed to only 0.75–1.0 $mg \cdot L^{-1}$ Ca. Aluminum levels $\geq 50$ $\mu g \cdot L^{-1}$ enhanced survival of fish exposed to pH 4.8, regardless of Ca level, but the enhancement was accentuated with increasing Ca levels. Almost no fish survived when exposed to pH $\leq 4.4$, regardless of Al and Ca levels (Ingersoll et al. 1990a; Wood et al. 1990a).

There are fewer studies on adult brook trout responses to dissolved Al than on alevins and juveniles, but adults may be as sensitive as younger fish. Survival decreased after 54 d of exposure to 77–156 $\mu g \cdot L^{-1}$ Al at pH 5.0 (Mount et al. 1988) and after 28 d of exposure to 118–296 $\mu g \cdot L^{-1}$ and pH 4.4–5.4 (Ingersoll et al. 1990b). Aluminum levels affecting electrolyte control or blood parameters in adults were similar to those decreasing survival. However, the responses of electrolytes and other blood parameters occurred within a few hours or days although survival did not decrease until 10 d. Net Na losses were reported at 111–333 $\mu g \cdot L^{-1}$ Al and pH 4.8–5.2 (Booth et al. 1988; McDonald et al. 1991). Hematocrit increased and blood pH decreased at the same ranges (Table 8) (Walker et al. 1991; Wood et al. 1988a–c). There have not been sufficient studies to make generalizations about metal uptake by adult brook trout.

**Lake Trout** (*Salmo namaycush*). Lake trout occur throughout most of North America from the Great Lakes drainage northward and in isolated mountainous areas of the Mississippi River drainage. Introduced populations also occur in the upper Colorado River drainage and some west coast drainages (Lee et al. 1980). Although many of these areas receive acid deposition, few studies have been conducted on the species' Al sensitivity; most of these have concentrated on alevins and juveniles.

Short-term exposure of alevin and juvenile lake trout to pH 4.4–5.1 and 50–300 $\mu g \cdot L^{-1}$ Al causes adverse physiological effects and reduces survival (see Tables 7 and 8). Alevins exposed to pH near 5.0 and 101–199 $\mu g \cdot L^{-1}$ Al for 5 d followed by 21-d and 32-d recovery periods had reduced whole-body electrolyte levels and growth (Gunn and Noakes 1987). Alevins may be more sensitive than juveniles to Al at low pH. Although alevin survival was reduced with exposure for 13 d to both pH 4.4 and 4.6 and 50 $\mu g \cdot L^{-1}$ Al, juvenile survival was reduced at 50 $\mu g \cdot L^{-1}$ and pH 4.4 but not at pH 4.6 until Al exceeded 300 $\mu g \cdot L^{-1}$. Levels of 50–300 $\mu g \cdot L^{-1}$ did not enhance alevin or juvenile survival during exposure to pH 5.0 or lower (Hutchinson et al. 1989).

**Fathead Minnow** (*Pimephales promelas*). The fathead minnow occurs in most drainages east of the continental divide from Chihuahua to Great Slave Lake but is absent from many coastal and Hudson Bay drainages. It has been introduced widely outside of the native range as a bait fish (Lee et al. 1980). The fathead minnow is commonly used in North America for laboratory studies.

Young fathead minnows are sensitive to low Al levels near their lower tolerance limits to pH. Hatching success and alevin survival were not reduced after several weeks exposure to pH 7.5 and 22–66 $\mu g \cdot L^{-1}$ Al. However, survival was significantly reduced with exposure to pH 6.0 and 25 $\mu g \cdot L^{-1}$ (McCormick et al. 1989). Although egg and alevin tolerance of low-pH water has not been studied, other results suggest that juvenile fathead minnows cannot tolerate pH of 5.5 or lower. Reduced survival occurred among juveniles exposed to pH 5.5 and 26 $\mu g \cdot L^{-1}$ Al for 96 hr, but no impact was observed at pH of 6.5 or higher and 50–400 $\mu g \cdot L^{-1}$ (see Table 7) (Palmer et al. 1988).

**White Sucker** (*Catostomus commersoni*). White suckers range from the Arctic to the southern U.S. They are absent from most Pacific drainages, many Arctic drainages between Hudson Bay and the McKenzie drainage, and Gulf Coast drainages (Lee et al. 1980). Their range includes most of the acid-vulnerable waters east of the Rocky Mountains (NAPAP 1991). Impacts of Al on white sucker survival and growth have been studied for the eyed egg through early juvenile life stages. Studies of metal accumulation have been conducted on wild-caught adults.

Higher percentages of eyed embryos than alevins survived when exposed to 50–350 $\mu g \cdot L^{-1}$ Al and pH 5.2–5.6 (see Table 7). Eyed egg but not alevin survival increased with 50–190 $\mu g \cdot L^{-1}$ at pH 5 compared to no addition of Al (Baker and Schofield 1982; Holtze and Hutchinson 1989).

Alevin survival was greatly reduced with Al (100–150 $\mu g \cdot L^{-1}$) at pH 5.2–5.6, and no alevins survived at pH below 5, regardless of Al levels. Alevins exposed to no Al and pH 7.1 reached about 20 g during the study, those exposed to pH 5.2–5.6 with no Al ranged from 14 to 19 g, and those surviving Al exposure were only about 4 g (Table 7) (Baker and Schofield 1982).

Juvenile white suckers appear to tolerate more Al than alevins at lower pH (Table 7). Aluminum above 85 $\mu g \cdot L^{-1}$ reduced survival of both age groups at pH below 5.6, but more juveniles than alevins survived at pH 5.0 and 90 $\mu g \cdot L^{-1}$ (Baker and Schofield 1982). Juvenile sensitivity also seems to diminish with size. At 60–100 $\mu g \cdot L^{-1}$ Al, the $LT_{50}$ of 14-mm juveniles was less than half that of 19-mm juveniles, but at higher levels the $LT_{50}$s of both size classes were about equal (Driscoll et al. 1980).

All studies of Al effects on white sucker survival and growth were performed at pH of 5.6 or lower. Thus, we cannot compare the sensitivity of white suckers to that of brook trout or rainbow trout whose tests occurred over a broader pH range. However, data on whole-body residues suggest that adult white suckers and trout may respond similarly to elevated Al at pH 5.6–7.0. In both groups, the greatest accumulation occurred near pH 6.0 (see Table 11). Concentrations in free-living adults at pH 5.8 water and 66 $\mu g \cdot L^{-1}$ Al was approximately twice that of those in pH 5.4 water and 163 $\mu g \cdot L^{-1}$ (Hamilton and Haines 1989).

**Striped Bass** (*Morone saxatilis*). Striped bass is an anadromous species whose native range includes coastal drainages from the St. Lawrence River to the St. Johns River, Florida, and from the Suwannee River, Florida, to Lake Ponchartrain, Louisiana. However, it has been introduced widely into lakes and impoundments throughout the U.S. and into San Francisco Bay on the west coast (Lee et al. 1980). Low recruitment in Chesapeake Bay led to studies of young striped bass susceptibility to reduced pH and elevated Al levels in the spawning areas of eastern Maryland streams. The pH sometimes drops below 6 in these streams during spring rainstorms (Hall et al. 1985). Only survival has been studied in this species.

Very young striped bass appear to be among the fish most sensitive to reduced pH and elevated Al; however, resistance appears to increase with age (Table 7). Survival of 11-d-old juveniles exposed to 25–200 $\mu g \cdot L^{-1}$ Al at pH 7.2 was 74%, but exposure to 400 $\mu g \cdot L^{-1}$ at pH 7.2 reduced survival significantly. Survival at pH 6.5 with no Al was similar to that at pH 7.2, but when 25 $\mu g \cdot L^{-1}$ Al was added, survival decreased during a 7-d test. Almost no 11-d-old fish survived pH 6.0 regardless of Al level. Almost all of the 160-d-old juveniles survived exposure at pH 7.2 and less than 400 $\mu g \cdot L^{-1}$ Al and at pH 6.5 and 200 $\mu g \cdot L^{-1}$ or less. Survival of 160-d-old juveniles at pH 6.0 and 50 $\mu g \cdot L^{-1}$ Al or more was significantly below that of pH 6.0 water with no Al (Buckler et al. 1987).

This study appeared to accurately reflect the fate of very young striped bass in acidified spawning habitats. Two 96-hr field toxicity studies were conducted at three Maryland striped bass spawning sites using newly hatched alevins. Survival among the acidified sites varied from 1% to 9.5%, and the average survival of unacidified was 84% (see Table 7). The pH at the spawning sites was comparable to those in the laboratory experiment, but Al levels were generally much higher (acidified, 150–2550 $\mu g \cdot L^{-1}$; unacidified, 0–520 $\mu g \cdot L^{-1}$) (Hall et al. 1985).

**Smallmouth Bass** (*Micropterus dolomieui*). Very little of the native range of smallmouth bass is vulnerable to acidification and Al mobilization. The original smallmouth bass range included the tributaries of the Mississippi River from the eastern Dakotas, Nebraska, and Kansas eastward. The species has been transplanted throughout the U.S. where there is suitable habitat (Lee et al. 1980).

Young smallmouth bass may be nearly as resistant to reduced pH and elevated Al as brook trout. Survival was 95% or greater among smallmouth bass exposed from 3 d before hatching to 4 d posthatch to pH 6.0–4.8 and 0–300 $\mu g \cdot L^{-1}$ Al and at pH 4.8 with 100 $\mu g \cdot L^{-1}$ or more Al. Survival was reduced at pH 4.5 and 0–50 $\mu g \cdot L^{-1}$, and at pH 4.2, regardless of Al level. Survival among fish exposed until 8 d posthatch was 96% or greater at pH 4.8 with 200 $\mu g \cdot L^{-1}$ (see Table 7) (Holtze and Hutchinson 1989). Survival was 80% or more among 48-hr-old alevins exposed for 96 hr to 1000 $\mu g \cdot L^{-1}$ and pH 7.5 and 6.1. However, no alevins survived 180 $\mu g \cdot L^{-1}$ or more at pH 5.1. In a 30-d exposure study with similarly aged fish, all alevins died at pH 5.5 and 200 $\mu g \cdot L^{-1}$ Al, and only 28% lived at pH 5.1 regardless of metal level (see Table 7) (Kane and Rabeni 1987). There was no evidence of reduced survival at pH near 6.1 from suffocation by precipitating Al as would be expected with the Neville and Campbell (1988) model.

Gills of smallmouth bass appear to accumulate greater Al levels than other tissues when exposed to pH about 6. Gill concentrations were approximately 15 times those measured in whole bodies of fish captured in the wild (see Table 11) (Brumbaugh and Kane 1985).

**Yellow Perch** (*Perca flavescens*) **and Walleye** (*Stizostedion vitreum*). Yellow perch occur in most Mississippi River drainages north of the Ohio River and in most Great Lakes and Hudson Bay drainages east of the Rocky Mountains. They occur in coastal drainages from South Carolina to New Brunswick and Nova Scotia. Walleye occur in most drainages east of the Rocky Mountains from the Arctic to headwaters of the Gulf Coast drainages. They are absent from the lower reaches of many Atlantic Coast drainages of the southeastern U.S. Both species have been introduced widely outside of their native ranges (Lee et al. 1980). Most of their ranges lie outside of acid-vulnerable areas of North America.

Survival studies with young walleye were conducted under pH and Al conditions similar to those used for smallmouth bass (Holtze and Hutchinson 1989). Walleye from the egg stage through 4 d posthatch may be more sensitive than alevins. Hatchling survival was reduced at pH 5.4 and 100 $\mu g \cdot L^{-1}$ Al compared to pH 5.4 with no Al, but was not reduced among alevins exposed to comparable Al levels and pH 4.0 and 4.7. Alevin, but not hatchling survival, seems to be enhanced in the presence of Al at pH 4.7 (see Table 7).

Studies of Al effects on yellow perch have only addressed residue concentrations in wild fish. Concentrations in gut tissues were more than 6 times those in the gills and more than 12 times greater than whole-body concentrations among fish collected in low pH (4.7–5.1) and circumneutral (pH 6.5) lakes. Gill tissue concentrations in fish from circumneutral lakes were generally higher than in fish from the lower-pH lakes. These results suggest that the concentrations of Al in food may have influenced those in the gut and in whole fish (see Table 10) (King et al. 1992).

### C. Influence of Dissolved Organic Carbon and Fluorine on Toxicity and Accumulation

Dissolved organic carbon (DOC) levels in natural waters generally vary from 1.2 $mg \cdot L^{-1}$ in clear waters, such as are found in the Adirondack region of New York, to 12 $mg \cdot L^{-1}$ in colored waters (Driscoll and Schecher 1990). Dissolved organic carbon can complex with 50%–70% of dissolved Al at pH 4.5 to 6.5, thereby reducing its availability to aquatic organisms.

Several investigations included exposures of Al with DOC levels that are typical of clear and colored water. The types of DOC used in these experiments have included refined pure organic compounds, such as citric acid, and commercially prepared organic mixtures prepared from soil humus. Most studies examined one or two levels of organic material to determine if DOC complexing reduces adverse effects of dissolved Al.

Levels of DOC up to 11–12 $mg \cdot L^{-1}$ appear to enhance survival of all life stages of Atlantic salmon, brook trout, and rainbow trout exposed to 500 $\mu g \cdot L^{-1}$ or less Al and pH 4.4 or lower (see Table 12). Citric acid increased brook trout alevin survival at 500 $\mu g \cdot L^{-1}$ Al and pH 5.5 (Driscoll et al. 1980), and humic material ($\leq$8.3 $mg \cdot L^{-1}$) increased brook trout juvenile survival at 340 $\mu g \cdot L^{-1}$ or less Al and pH 4.8–5.6 (Parkhurst et al. 1990). However, higher Al levels exceeded the protective effects of DOC. All Atlantic salmon alevins died when exposed to 1809 $\mu g \cdot L^{-1}$ Al (Peterson et al. 1989) and brook trout juveniles at 1174–1276 $\mu g \cdot L^{-1}$, regardless of DOC level (Parkhurst et al. 1990).

The ameliorative effect of DOC appears to be consistent among salmonids. Although Atlantic salmon parr were more vulnerable to Al (113 $\mu g \cdot L^{-1}$ at pH 4.9) than juvenile brook trout, the effects of DOC on survival were equally dramatic in both species (McCahon and Pascoe 1989).

Brouard et al. (1989) found that Al levels in DOC-exposed juvenile rainbow trout and embryonated Atlantic salmon eggs were about one-half that in unexposed fish (Table 13). Witters et al. (1990b) reported statistically significant reductions in Al levels among adult rainbow trout in the presence of DOC compared to fish in water without the mixture.

Dissolved F has a strong affinity for dissolved Al under acidic conditions. Because of its similar size, F readily substitutes for $OH^-$ at pH 4.3–7.3. However, F levels in natural waters are typically low (median = 190 $\mu g \cdot L^{-1}$) and at pH 5.5 or lower, Al levels often exceed F (Driscoll and Schecher 1990). Thus, in many environmental situations F does not ameliorate Al toxicity.

A few investigators have examined the potential impact of dissolved F on the survival of younger life stages of Atlantic salmon and brook trout. All F levels tested were within 2.5 times the median levels found in freshwater. Fish survival generally increased with F, reaching a maximum when F and Al levels were about equal. However, F was little help when Al levels were substantially higher (Parkhurst et al. 1990).

## D. Hazard Assessment

The hazard that dissolved Al poses to fish is influenced greatly by acidity, dissolved cations (particularly Ca), DOC, and, to a lesser extent, F. Stress caused by Al is influenced greatly by pH because of the intimate relationship between Al solubility and pH. Calcium tends to counteract Al by helping fish resist electrolyte loss across gill membranes. Both DOC and F tend to complex with monomeric Al, thereby reducing its availability as a stressing agent. Although $SO_4^{2-}$ also combines with monomeric Al, these complexes are weaker and less important than those with F except where $SO_4^{2-}$ is unusually high (Driscoll and Schecher 1990).

Acid-sensitive species are most likely to encounter toxic Al levels if they live in slightly acidic habitats (pH 5.8–7.0) receiving low pH, Al-laden runoff. In these habitats, monomeric Al can precipitate and settle on the gills, causing tissue irritation and reduced respiratory capacity (Skogheim et al. 1987). Blueback herring and fathead minnow, which are among the most acid-sensitive species studied, die when pH is less than 6.1. Other acid-sensitive species include redbelly dace (*Phoxinus* sp.), slimy sculpin (*Cottus cognatus*), blacknose dace (*Phoxinus cumberlandensis*), bluntnose minnow (*Pimophales notatus*), blacknose shiner (*Notropis heterolepis*), common shiner (*Notemigonus cornutus*), fallfish (*Semotilus corporalis*), and pearl dace (*Semotilus margarita*) (Baker and Christensen 1991; Pauwels and Haines 1986).

Fish that can tolerate pH of 5.8 or less are most likely to encounter dangerous Al levels in acidic habitats (i.e., pH 4.2–5.8) and in slightly acidic habitats receiving low pH, Al-laden runoff. Most of the species discussed in this review fall within this category. Other species with pH limits in this range include mudminnow (*Umbra* sp.), northern pike (*Esox lucius*), golden shiner (*Notemigonus crysoleucas*), creek chub (*Semotilus atromaculatus*), finescale dace (*Phoxinus neogaeus*), brown bullhead (*Ictalurus nebulosus*), channel catfish (*Ictalurus punctatus*), rock bass (*Ambloplites rupestris*), pumpkinseed sunfish (*Lepomis gibbosus*), and bluegill sunfish (*Lepomis macrochirus*) (Baker and Christensen 1991; Berg and Burns 1985; Palmer et al. 1988; Pauwels and Haines 1986).

Fish in waters mitigated to control acidity also can be exposed to dangerous Al levels. Some mitigation efforts involve injecting slurries of pulverized limestone ($CaCO_3$) into a stream; others involve inserting containers of a $CaCO_3$ source, such as limestone cobble or mixtures of clam shells and sand, into acidified streams. These measures can raise the pH from the mid-4.0s to 5.8–7.0 in a narrow mixing zone. The sudden rise in pH with the concomitant drop in Al solubility can cause Al precipitation and fish toxicity (Rosseland and Skogheim 1986; Rosseland et al. 1992).

The greatest hazard to fish populations in acidified environments is reduced or failed recruitment (Baker and Schofield 1982; Haines 1981). Thus, most studies of Al toxicity have targeted the youngest life stages. However, studies involving more than one life stage sometimes have produced inconsistent results about the relative vulnerability of life stages in a given species or among species. Sensitivity among cutthroat trout and Atlantic salmon appears to follow the sequence juveniles > alevins > embryos (Skogheim and Rosseland 1984, 1986; Woodward et al. 1989). In one study on brook

Table 13. Influence of DOC ($mg \cdot L^{-1}$) on Al concentrations ($\mu g \cdot g^{-1}$) in fish experimentally exposed to different levels of Ca ($mg \cdot g^{-1}$), Al ($\mu g \cdot L^{-1}$), and pH.

| Species | Experimental conditions | | | | | | | | | Tissue Al[1] | Reference |
|---|---|---|---|---|---|---|---|---|---|---|---|
| | Life stage | Temp (°C) | pH | Total Al | Available Al | $Ca^{2+}$ | DOC | Organic material | Exposure time | | |
| *Oncorhynchus mykiss* | Juvenile | 0.5 | 5.5 | 600 | | | 0 | humic material | 7 d | 12.5 | 1 |
| | | | | | | | 10 | | | 5.7 | |
| *Oncorhynchus mykiss* | adult | 8–10 | 4.7 | 179 | 175 | 3.5 | 0 | | 10 d | 75.0 | 2 |
| | | | | 184 | 36 | | 10 | | | 4.5 | |
| *Salmo salar* | eggs | 0.5 | 5.5 | 300 | | | 0 | | 14 d | 49 | 1 |
| | | | | | | | 10 | | | 29 | |
| | | | | | | | 0 | | | 15.5 | |
| | | | | | | | 10 | | | 5.4 | |
| *Salmo salar* | alevin | 6.0 | 5.0 | 0 | | 0.4 | 0.25 | citric acid | 43 d | 67 | 3 |
| | | | | | | | 2.50 | | | 66 | |
| | | | | 181 | | | 0 | | | 204 | |
| | | | | | | | 0.25 | | | 150 | |
| | | | | | | | 2.50 | | | 85 | |
| | | | | 1809 | | | 0 | | | 163 | |
| | | | | | | | 0.25 | | | 150 | |

| | | | | | | | | | | |
|---|---|---|---|---|---|---|---|---|---|---|
| *Salmo trutta* | adult | 2.5 | 5.5 | 50 | 6.0 | 0 | humus | 42 d | 65 | 4 |
| | | | | | | 20[2] | | | 17 | |
| | | | | 200 | | 0 | | | 140 | |
| | | | | | | 20 | | | 90 | |
| | | | | 500 | | 0 | | | 95 | |
| | | | | | | 20 | | | 130 | |
| | | 15 | | 50 | | 0 | | | 36 | |
| | | | | | | 20 | | | 18 | |
| | | | | 200 | | 0 | | | 67 | |
| | | | | | | 20 | | | 66 | |
| | | | | 500 | | 0 | | | 97 | |
| | | | | | | 20 | | | 160 | |

1. Wet weight in rainbow trout gill and Atlantic salmon eggs (Brouard et al. 1989); wet weight in rainbow trout gill (Witters et al. 1990b); dry weight, whole body in Atlantic salmon alevin (Peterson et al. 1989); and dry weight in brown trout gill (Karlsson-Norrgren et al. 1986b). All concentrations reported in $\mu g \cdot g^{-1}$.

2. Concentrations measured in platinum units.

References: (1) Brouard et al. 1989; (2) Witters et al. 1990b; (3) Peterson et al. 1989; (4) Karlsson-Norrgren et al. 1986b.

trout, the sequence of sensitivity was juveniles > alevins > embryos (Baker and Schofield 1982), whereas in other studies it was juveniles = alevins > embryos (Cleveland et al. 1986, 1989; Ingersoll et al. 1990a).

Sensitivity in alevin and juvenile lake trout appears to be about equal (Hutchinson et al. 1989). In fathead minnows, the sequence is juveniles > alevins = embryos (McCormick et al. 1989). In white suckers, the sequence was juveniles > alevins ≥ embryos (Baker and Schofield 1982; Holtze and Hutchinson 1989). However, among striped bass, older juveniles were clearly less sensitive than very young juveniles (Buckler et al. 1987). Among walleye, embryos appeared to be more sensitive than alevins (Holtze and Hutchinson 1989). It seems reasonably safe, however, to conclude that adult fish are less sensitive to Al or pH than young fish of the same species. Thus, these distinctions have little practical significance. Resource managers probably should direct their habitat management efforts toward protecting important species from the late egg stage through the first several weeks of the juvenile stage.

Aluminum-induced stress tends to be additive to that caused by declining pH. Levels of 100–500 $\mu g \cdot L^{-1}$ commonly are needed to cause death or a physiological response when pH is above tolerance limits. However, levels below 100 $\mu g \cdot L^{-1}$ may cause death or a physiological response when a species is near its pH limits. At harmful levels of pH, acidity masks toxic effects of Al.

Some studies have indicated that Al may ameliorate the toxic effects of acidity under laboratory conditions. Usually, the ameliorative effect raises survival rates a few percentage points or for a few hours. These ameliorative effects have not been reported under field conditions, probably because most populations are exposed for longer periods of time or sampling procedures are not sufficiently accurate to measure any effect. Thus, it would be unwise for resource managers to count on elevated Al to reduce $H^+$ stress on wild fish populations.

Experimental levels of Ca typically vary from less than 1.0 to 38 $mg \cdot L^{-1}$, a range that generally falls within that of natural, poorly buffered water bodies in Europe and North America. Several studies have shown that small (1–12 $mg \cdot L^{-1}$) increases in Ca levels can dramatically improve alevin and adult survival in brown trout, brook trout, and rainbow trout when exposed to elevated Al and $H^+$ under laboratory conditions (Booth et al. 1988; Brown 1983; Playle et al. 1989; Sadler and Lynam 1988; Sayer et al. 1991b). The exposure times for these studies varied from 66 hr to 10 wk. These results suggest that Ca may be important in enhancing fish survival during brief periods of reduced pH and elevated Al, such as during spring snow melt. Aluminum toxicity can also be ameliorated by DOC and F.

Field studies evaluating the potential impacts of several environmental factors suggest that Al usually plays a subordinate role in determining the composition of fish communities. Pauwels and Haines (1986) studied relationships between fish communities and nine chemical and morphological characteristics of 22 lakes in Maine (pH 4.4–7.0). Through a stepwise multiple regression analysis, six of the nine variables accounted for 86% of the accountable variance, with divalent cation level accounting for 58% of the variance. Only 3% of the accountable variance was due to Al.

Matuszek and Beggs (1988) examined 19 physical and chemical parameters and fish communities of 992 Ontario lakes. Of these, 272 lakes had pH below 6.0. Their overall analysis revealed that lake area (18%), latitude (6%), Al (5%), and DOC (3%) accounted for the greatest amount of variance in fish communities. When the authors considered only lakes with pH less than 6.0, they showed that acidity was the dominant factor and accounted for 21% of the variance. The authors did not attempt to distinguish between pH and Al influences in these acidic lakes.

## VI. Effects on Amphibians and Reptiles

### A. Amphibians

As with fish, most research on the toxicological effects of Al on amphibians has occurred in conjunction with acid deposition. Low pH is toxic to amphibians (Freda 1986), and it is often difficult to distinguish effects of Al from those caused by acidity.

Freda (1991) pointed out that Al toxicity to amphibians is dependent on species, stage of development, pH, and chemical and physical composition of the factors in the water, previous exposure, and genetic variation within a species. The $LC_{50}$ in the leopard frog (*Rana pipiens*) at pH 4.8 was between 403 and 471 $\mu g \cdot L^{-1}$, but it was 746 $\mu g \cdot L^{-1}$ in the wood frog (*Rana sylvatica*) at the same pH (Freda and McDonald 1990). Similarly, older tadpoles of the common frog (*Rana temporaria*) were 2.5 times more tolerant of Al at pH 4.4 than were American toad (*Bufo americanus*) tadpoles (Freda 1991).

Newly hatched tadpoles tend to be the most sensitive life stage, followed by embryos and older tadpoles. For example, $LC_{50}$s for leopard frog embryos, young tadpoles, and older tadpoles at pH 4.8 were 403, less than 250, and greater than 1000 $\mu g \cdot L^{-1}$ Al, respectively (Freda and McDonald 1990). An exception to this occurred in the American toad, for which no toxicity could be detected in young embryos and newly hatched tadpoles but was obvious in older embryos (Clark and LaZerte 1985; Freda et al. 1990).

Freda (1991) recognized three types of responses to Al and pH displayed by amphibian embryos and larvae. In the Type I response, demonstrated by American toads, spring peepers (*Hyla crucifer*), and embryos of wood frogs, amphibians are only sensitive to Al at pH less than 4.5. In the Type II response, as displayed by some leopard frogs and wood frogs, embryos only become sensitive to Al when pH is greater than 4.5. Mortality below this pH can be explained by acidity alone. In the Type III response, Al is toxic at pH above 4.5 but ameliorates the toxic effects of acidity at lower pH. This response has been seen in spotted salamanders (*Ambystoma maculatum*) and in some studies with leopard frogs (Freda and McDonald 1990).

In a study that combined field observations with laboratory experimentation, Horne and Dunson (1994) found that Al and $H^+$ were entirely confounded with each other but were consistently lower in Pennsylvanian ponds that contained larval Jefferson salamanders (*Ambystoma jeffersonianum*) than in those that did not have the salamander. No ponds with pH below 4.4 contained Jefferson salamanders. In the laboratory, the researchers found that Al greater than 250 $\mu g \cdot L^{-1}$ ameliorated the toxic effects of pH 4.5, with maximal benefit at 490 $\mu g \cdot L^{-1}$.

No differences in toxicity among the monomeric or soluble forms of Al have been found for amphibians (Clark and LaZerte 1985; Freda 1991). As with fish, DOC may bind with Al and make it biologically unavailable (Freda et al. 1990). However, DOC greater than 12 $mg \cdot L^{-1}$ may be toxic to amphibians (Clark and Hall 1985; Freda et al. 1990) and eliminate any ameliorative benefits.

The most overt response by amphibian embryos to acid or Al toxicity is curling of the tail, which may result from the inflexibility of the vitelline membrane surrounding the embryo or malfunctions in the production or action of hatching enzymes. Hatching is often reduced under these conditions, and those tadpoles that do hatch may have severe bending of the spine or scoliosis. Cummins (1986) determined that 800 $\mu g \cdot L^{-1}$ Al at pH 4.4 retarded growth and metamorphosis of common frog tadpoles, but pH 4.4 without added Al had no effect. Twice that amount of the metal stunted tadpoles and eventually killed small ones, although larger ones metamorphosed. Newly hatched tadpoles may develop epithelial fluid-filled cysts in waters with high Al, indicative of skin irritation (Freda and McDonald 1990). Older tadpoles exposed to elevated Al levels lose plasma electrolytes, particularly Na and Cl, which suggests that toxicity is manifested by failures in osmoregulation (Freda and McDonald 1990). Freda (1991) suggested that the gelatinous mass surrounding amphibian eggs may afford some protection against dissolved Al, which could account for the reduced sensitivity of embryos compared to newly hatched larvae.

The risk an amphibian faces from Al toxicity is, in part, related to its habitat. Freda (1991) distinguished five risk categories based on breeding areas:

1. Temporary ponds: small (<20 m diam), shallow (<1 m avg depth) pools subject to drying in midsummer. The only sources of water for these ponds are rainfall and snowmelt. These ponds are susceptible to acidification and elevated Al, especially with spring snow melt (Schaefer et al. 1990), when breeding is at its peak for many amphibian species. Approximately 30% of all salamander and 50% of all frog species in North America breed in temporary ponds. Common amphibians in this

category include salamanders (*Ambystoma* spp., *Notopthalmus* spp.), treefrogs (Hylidae), toads (*Bufo* spp.), and wood frogs.

2. Lakes, rivers, and permanent ponds: permanent bodies of water that are less vulnerable to wide fluctuations in chemistry or physical parameters because of their size and depth. These bodies of water can be affected by perturbations in their watersheds such as acid mine drainage or acid precipitation, especially when surrounding soils are poorly buffered. Common amphibians breeding in these waters include large frogs such as bullfrogs (*Rana catesbieana*) and green frogs (*Rana clamitans*), which breed throughout the summer and do not concentrate their breeding in the potentially more hazardous early spring.
3. Bogs and fens: waters that are acidified naturally through *Sphagnum* moss and organic acids. Bogs are isolated, whereas fens receive some of their water and minerals from external sources. Because of their acidity, bogs and fens may have elevated levels of Al and other metals, but they are typically bound to organic molecules. High DOC and low pH may make these areas unsuitable for most amphibians.
4. Small streams: mountain headwater streams and seeps that are vulnerable to pulses of acid and Al in the spring. Plethodontid salamanders are common residents of these areas and may be adversely affected by episodic events of metal or acid increases.
5. Soils: A few amphibians, such as red-backed salamanders (*Plethodon tristatus*), are entirely terrestrial, and some adult ambystomid salamanders also live underground for most of their life. Reduced soil pH caused by acid precipitation or acid mine drainage and concomitant elevation of dissolved Al may be hazardous to these species (Wyman and Jancola 1992).

We were unable to find any published study that reported the level of Al in tadpoles. However, Sparling and Lowe (1995) analyzed whole bodies of *Acris crepitans* and *Hyla versicolor* collected from experimentally treated (acidified to pH 4.8 with or without added Al) macrocosms. They found that mean Al levels in *Acris crepitans* ranged from 10,620 to 16,920 $\mu g \cdot g^{-1}$ dry wt and were significantly higher than those in *Hyla crucifer* (range 3460–11,470 $\mu g \cdot g^{-1}$). Added Al did not affect Al levels in the bodies of these species, but DOC was sufficiently high ($>5$ $\mu g \cdot L^{-1}$) that much of the aqueous Al probably was bound and biologically unavailable. Concentrations of Al in gut coils of *Rana clamitans* ($\bar{x}$ = 20,840 ± 5900 $\mu g \cdot g^{-1}$) from the same experimental complex were significantly greater than those in body tissues ($\bar{x}$ = 634 ± 416 $\mu g \cdot g^{-1}$) but undistinguishable from those of the sediments ($\bar{x}$ = 15,720 ± 6870 $\mu g \cdot g^{-1}$).

In summary, information on Al toxicity in amphibians is incomplete, sometimes contradictory, and confounded by the toxicity of low pH. As a general rule, elevated levels of Al are likely to be toxic when pH drops below 4.5, although some species show signs of toxicity within the pH range of 4.5–5.5. Lethal levels may be as low as 150 $\mu g \cdot L^{-1}$, but $LC_{50}$s are commonly several times higher. Early life stages, especially newly hatched tadpoles, are most sensitive. Because many species of amphibians breed in early spring, episodic events of reduced pH and elevated Al are particularly important to this group.

### B. Reptiles

No scientific papers have been published on the effects of Al on reptiles. Presumably, aquatic and fossorial reptiles may be impacted by ambient or dietary Al, especially in acidic environments. Risk may be reduced compared to amphibians and freshwater fishes because even aquatic reptiles nest on land. However, survivorship of eggs may be affected by high levels of Al in acidic soils. Research is required on possible risk to species such as turtles and water snakes.

### C. Hazard Assessment

Amphibian populations are at elevated risk from Al under acidic conditions, but high $H^+$ is also toxic and it is difficult to distinguish the effects of the two factors. Ambient levels of soluble Al are acutely toxic at 250 $\mu g \cdot L^{-1}$ for some species, and physiological problems appear below 100 $\mu g \cdot L^{-1}$. The species at greatest risk are those that inhabit small temporary or ephemeral ponds that receive most of their water from spring runoff and snow melt. Aluminum levels and acidity in these waters are particularly high, and exposure occurs during early life stages when amphibians tend to be most sensitive. In North America, a large proportion of amphibian species breed in these conditions. Larger species that inhabit lakes and permanent ponds may be less at risk from episodic events but can still be affected by chronic acidification and accompanying elevated Al, especially where soils are poorly buffered. Tadpoles may have very high levels of Al. Depending on speciation and bioavailablity of Al in the bodies and guts of these organisms, they may provide a risk of secondary toxicity to predators. Risks to most reptiles appear minimal, but there are no data on aquatic or semiaquatic species.

## VII. Effects on Birds

Birds are most likely to be exposed to Al through their diets. Fortunately, only a fraction of the ingested metal is assimilated, and fecal excretion is relatively efficient. Thus, birds are less at risk than fish, amphibians, or aquatic invertebrates.

### A. Physiological Effects

In general, diets with less than 1000 $mg \cdot kg^{-1}$ Al are not considered harmful. Bones, particularly in growing chicks and reproductive females, appear to be the most efficient tissues for storing the metal, followed by brains, livers, and kidneys. Levels in feathers of raptor chicks may also reflect dietary exposure (Bortolotti and Barlow 1988), but only one species has so far been examined. Scheuhammer (1991) and Scanlon (1990) have reviewed the toxicology of Al in birds.

Aluminum has not been identified as a required mineral. However, indirect evidence (National Academy of Sciences 1980) suggests that low dietary levels may be necessary. Toxic effects of high dietary Al typically are associated with problems in Ca and P metabolism. The metal binds with P in the gut to form an insoluble complex and decreases the availability of P in the diet. Signs of toxicity, therefore, are consistent with P deficiency and include reduced growth, impaired eggshell quality, loss of appetite, decreased laying, feather molt, and rickets. Harmful effects are most apparent when diets are deficient in P or when Al levels exceed 50% of P. Supplementary P can ameliorate toxicity (Scheuhammer 1991).

Under laboratory conditions, Hussein et al. (1988) found that diets with 3000 $mg \cdot kg^{-1}$ Al (dry wt), 3500 $mg \cdot kg^{-1}$ P, and 25,000 $mg \cdot kg^{-1}$ Ca depressed food consumption in Japanese quail (*Coturnix c. japonica*) beginning with the first day of treatment. By 4 d of treatment, egg production had dropped by 90%. Eggshell breaking strength and body weight also declined compared to control birds. Dietary Al levels of 1500 $mg \cdot kg^{-1}$ temporarily reduced food intake and eggshell breaking strength and decreased egg production and body weight during the 4-wk study.

When domestic white leghorn hens were fed graded levels of Al (0–3000 $mg \cdot kg^{-1}$), 5000 $mg \cdot kg^{-1}$ P, and 33,000 $mg \cdot kg^{-1}$ Ca, those on the highest Al levels reduced egg production and feed intake during the first half of a 6-wk study (Hussein et al. 1989). Egg production and food consumption rose to control levels during the second half of the study. Mean body weight varied inversely with Al level, and plasma P declined at or above 1000 $mg \cdot kg^{-1}$ Al.

Tibial breaking strength was significantly reduced at 2000 $mg \cdot kg^{-1}$ Al compared to controls but not at 3000 $mg \cdot kg^{-1}$. This may be due to greater food consumption and hence higher levels of metal uptake by hens on 2000 $mg \cdot kg^{-1}$ Al. Similar responses in reduced food consumption, growth, and plasma P were seen in chicks fed diets with 3000 $mg \cdot kg^{-1}$ Al (Wisser et al. 1990). Reduced egg production probably results from Al interfering with ATPase activity of the eggshell gland mucosa (Lundholm and Mathson 1986).

Chicken embryos that were inoculated with as little as 15 mg Al as $AlCl_3$ at 3 d of incubation developed several malformations by 9 d of incubation, including reduced body size, microphthalmia, shortened limbs, ectopic viscera, microcaudia, and hemorrhages (Gilani and Chatzinoff 1981). The lowest dose of $AlCl_3$ (3 mg/egg) reduced embryo survival by 31%, and 18 mg $AlCl_3$ lowered it by 72%. Firling et al. (1994) determined that a single 100-$\mu$L dose or repeated 25-$\mu$L doses of aluminum citrate in chicken embryos resulted in deformed and shortened tibias but did not influence mortality or body weight.

Toxicity was age dependent in capercaillie (*Tetrao urogallus*) chicks (Spidso and Staurnes 1991). When 4000 $mg \cdot kg^{-1}$ Al was added to commercial poultry food, 1-wk-old chicks experienced 100% mortality in 14 d compared to 30% mortality in chicks fed 400 $mg \cdot kg^{-1}$ and 20% in control birds. However, 4000 $mg \cdot kg^{-1}$ Al did not increase mortality in 5-wk-old chicks during 3 wk of exposure. The older chicks also had higher plasma Ca than young birds and measurable levels of Al in kidney and liver.

Concern about ecological consequences of acid precipitation has led to a few field studies dealing with the effects of aqueous Al on birds. In Sweden, 30%–40% of pied flycatcher (*Ficedula hypoleuca*) nests from a population near an acidic lake had impaired eggs (Nyholm 1981; Nyholm and Myhrberg 1977). Deformities included soft spots caused by defective palisade layers in the shell. Porosity within these soft spots was greater than in normal eggs and led to excessive dehydration and reduced hatchability. Defective eggs were typically among the last laid in a clutch and only occurred in nests within 25 m of the water's edge. Levels of several possible contaminants and metals (DDT, PCB, Al, Cd, Cr, Cu, Pb, Hg) were measured, but only Al levels differed between these birds and those living farther from the lake. Female flycatchers had elevated levels of Al and abnormal apatite in their humeri. Nyholm (1981) hypothesized that elevated Al levels in invertebrates within the lake became toxic. He later reported levels up to 1230 $mg \cdot kg^{-1}$ in aquatic invertebrates from the lake but did not report Ca or P levels.

Nyholm's (1981) hypothesis stimulated interest in a possible role of Al in decreasing avian recruitment, but few studies have substantiated his findings. In Ontario, water chemistry parameters including Al concentrations accounted for a small but significant percentage of the total variance in egg water loss, egg volume, and growth of embryos (Glooschenko et al. 1986). However, variation among nests on a lake and among siblings within a nest accounted for greater amounts of variance.

Carriere et al. (1986) tested Nyholm's hypothesis of Al toxicity by exposing breeding pairs of ringed turtle doves (*Streptopelia risoria*) to 1000 $mg \cdot kg^{-1}$ dietary Al as $Al_2(SO_4)_3$. Both Ca (9000 $mg \cdot kg^{-1}$) and P (5000 $mg \cdot kg^{-1}$) levels were below recommended dietary levels, although not as low as that found in many species of aquatic insects (Sadler and Lynam 1985). No differences were observed in breeding success or eggshell structure as a result of treatment. Although femurs from females on the Al-supplemented diets had twice the concentration of Al as those from controls, there was no difference in plasma Ca, P, or Mg levels.

The researchers also fed 21-d-old chicks diets with graded Al levels (0–1500 $mg \cdot kg^{-1}$). No differences were detected in growth rates or in tissue levels of Ca or P. Aluminum levels in sterna of chicks on 1500 $mg \cdot kg^{-1}$ were higher than those at lower concentrations.

Dippers (*Cinclus cinclus*) have been studied in Scotland and Wales by Ormerod et al. (1988), who found that aqueous Al concentrations correlated inversely with stream pH whereas aqueous Ca increased with pH. However, they could not correlate aqueous Al levels with those in food organisms. Although thickness of dipper eggshells declined with pH, there was no correlation between eggshell thickness and Al in either insect prey or water.

Livers of goldeneyes (*Bucephula clangula*) in Sweden were examined for metal content related to acidification and liming of lakes (Eriksson et al. 1989). Limed lakes actually had higher concentrations of aqueous Al than acidified or circumneutral lakes, although the levels were not statistically distinguishable. No differences in liver concentrations were found for any metal (Table 14).

Grue et al. (unpublished data) fed adult European starlings (*Sturnus vulgaris*) diets varying in levels of Ca, P, and Al to determine if the metal interacted with the other elements to affect reproductive capability. Starlings on diets with elevated Al (1000 or 5000 $mg \cdot kg^{-1}$) but normal levels of Ca and P laid significantly more eggs than those on control levels (200 $mg \cdot kg^{-1}$). Low levels of Ca, however, reduced clutch sizes. Diets with low Ca and low P reduced food consumption and body weights of the starlings. Elevated Al may have increased mortality in adults when combined with low or normal levels of P, but the differences were not statistically significant. Miles et al. (1993) also determined that eggshells of starlings on elevated Al were thicker, but not stronger, than those on low-metal diets but that egg and bone characteristics were affected more by dietary Ca and P levels than by Al.

Support for Nyholm's hypothesis was presented by Wolff and Phillips (1990), who fed Japanese quail regimens that had either low Ca (6500 $mg \cdot kg^{-1}$) and low P (4000 $mg \cdot kg^{-1}$) or high Ca (12,000 $mg \cdot kg^{-1}$) and high P (6000 $mg \cdot kg^{-1}$); each regimen was further divided into diets with 0, 120, 500, or 1500 $mg \cdot kg^{-1}$ Al as aluminum citrate. The low-Ca, low-P regimen caused eggshell thinning when compared to the high-mineral regimen but, within the low-Ca and low-P regimen, 120 $mg \cdot kg^{-1}$ Al further reduced eggshell thickness. Specific gravity, which often correlates with breaking strength, also declined with Al level, especially in the low-Ca and low-P, 1500 $mg \cdot kg^{-1}$ diet. For the high-Ca and high-P regimen, the percentage of broken eggs increased with elevated dietary Al levels even though there were no differences across diets in the number of eggs laid.

Sparling (1990) found evidence for Al toxicity in young waterfowl. He fed 1-d-old black duck (*Anas rubripes*) and mallard (*Anas platyrhynchos*) ducklings one of three base regimens: low Ca (3600 $mg \cdot kg^{-1}$) and low P (6200 $mg \cdot kg^{-1}$) [LL]; normal levels of Ca (15,100 $mg \cdot kg^{-1}$) and P (13,500 $mg \cdot kg^{-1}$) [NN]; and low Ca, high P (21,500 $mg \cdot kg^{-1}$) [LH]. Each regimen was supplemented with one of three Al levels (200, 1000, and 5000 $mg \cdot kg^{-1}$). Ducklings on the LL regimen were most affected by dietary Al;

Table 14. Aluminum concentrations ($mg \cdot kg^{-1}$) in tissues from free-ranging and experimental birds.

| Species | Tissue | Treated | Control | Remarks | Reference |
|---|---|---|---|---|---|
| *Melanocitta perspicillata* | liver | | 3.5–5.1 | hatch-year birds, Oregon and Washington | 1 |
| | liver | | 3.2–4.0 | adults, Oregon and Washington | 1 |
| | liver | | 5.2–12.6 | adult males, San Francisco Bay, CA | 2 |
| *Anas rubripes* | femur | 41.5–120 | 14.0–19.6 | ducklings fed 200 or 5000 $mg \cdot kg^{-1}$ as $Al_2(SO_4)_3$ for 2 wk | 3 |
| | femur | 19.5–20.5 | 5.4–7.4 | as above but fed for 10 wk | 3 |
| | liver | 8.0 | 0.4–0.8 | fed for 10 wk | 3 |
| *Anas platyrhynchos* | femur | 53.5–150 | 12.5–17.9 | as above but fed for 2 wk | 3 |
| | femur | 18.0–23.0 | 4.5–9.5 | fed for 10 wk | 3 |
| | liver | 8.5 | 1.6–4.3 | fed for 10 wk | 3 |
| *Bucephula clangula* | liver | 0.4–1.0[a] | | free-ranging ducklings | 4 |
| *Haliaeetus leucocephalus* | back feathers | 8.5–27.8 | | wild nestlings | 5 |
| | tail feathers | 77.5–175 | | wild nestlings | 5 |
| *Canachites canadensis* | feathers | 8.5–658 | | free-ranging | 6 |
| *Gallus domestica* | bone | 9.7 | 1.5 | chicks fed 3000 $mg \cdot kg^{-1}$ Al as $Al_2(SO_4)_3$ | 7 |
| | bone | 19.7 | 8.3 | adults fed 3000 $mg \cdot kg^{-1}$ as $Al_2(SO_4)_3$ | 7 |
| | egg | 11.8 | 11.9 | as above | 7 |
| *Streptopelia risoria* | femur | 7.4–7.9 | 6.7–15.9 | fed 1000 $mg \cdot kg^{-1}$ Al as $Al_2(SO_4)_3$ | 8 |
| | kidney | 0.9–1.9 | 0.7–1.0 | as above | 8 |
| | sternum | 9.9 | 15.9 | fed 1500 $mg \cdot kg^{-1}$ Al as $Al_2(SO_4)_3$ | 8 |
| | leg and wing bone | 6.6 | 6.7 | as above | 8 |
| *Tyrannus tyrannus* | eggshell | 0.42–0.33 | | free-ranging | 9 |
| *Sturnus vulgaris* | eggshell | 31–70 | 27–32 | fed 200 or 5000 $mg \cdot kg^{-1}$ as $Al_2(SO_4)_3$ | 10 |

[a]Concentrations expressed as fresh weight; other values as dry wt.

References: (1) Henny et al. 1991; (2) Ohlendorf et al. 1991; (3) Sparling 1991; (4) Eriksson et al. 1989; (5) Bortolotti and Barlow 1988; (6) Bortolotti et al. 1988; (7) Wisser et al. 1990; (8) Carriere et al. 1986; (9) Glooschenko et al. 1986; (10) Miles et al. 1993.

all birds on 5000 $mg \cdot kg^{-1}$ died within 2 wk, and mortality exceeded 60% for those on 1000 $mg \cdot kg^{-1}$. In addition, growth rates were reduced in black ducks on all diets except NN: 1000 $mg \cdot kg^{-1}$ compared to controls. Within each regimen, growth was inversely related to dietary Al level. Reduced growth rates were partially explained by inappetence in that food consumption was reduced in both mallards and black ducks on LL: 1000 $mg \cdot kg^{-1}$, and in black ducks on NN and LH with 5000 $mg \cdot kg^{-1}$ Al.

After 10 wk of treatment, there were strong correlations between Al levels in femurs and livers with Al in the diets (Sparling 1991). Black duck and mallard femurs from the NN regimen were stronger and heavier than those from either of the other regimens, suggesting that dietary Ca was more important than Al in determining bone structure. Breaking strength of black duck femurs did decline with the highest levels of Al within a regimen.

Clinical signs caused by Al or reduced Ca included lameness, discoloration, or mottling of the bill associated with a leathery texture of the epithelium of the upper mandible, complete fractures of a wing or leg, greenstick (rehealed) fractures involving a bending of the femur or tibiotarsus, and extremely flexible or rubbery femurs. Mortality and clinical signs were more widespread in black ducks than in mallards (Sparling 1990).

Capdevielle and coworkers compared the effects of Al, acid, and sulfate on growing domestic chickens and mallard ducklings in a well-controlled study including groups that were either pair-fed or fed *ad libitum*. In chickens, 5000 $mg \cdot kg^{-1}$ dietary Al as $Al_2(SO_4)_3$ reduced growth by 50% compared to *ad libitum* controls and 45% compared to pair-fed controls (Capdevielle and Scanes, 1995a). The high-Al diet also caused a 25% mortality in chicks, produced morbidity in survivors, reduced insulin-like growth factor, and lowered food intake. Liver levels of Al were positively related to dietary concentrations. Femurs and tibia of chicks fed high Al were soft, pliable, and had lower levels of ash, Ca, and P than controls. The soft bones may have resulted from depressed serum vitamin D (Capdevielle et al. in manuscript). Additions of sulfuric acid without Al also depressed growth compared to *ad libitum* controls but not to pair-fed birds, suggesting that the effects were largely the result of reduced food intake.

Fewer differences due to Al were observed in mallards (Capdevielle and Scanes, 1995b). Although ducklings on 5000 $mg \cdot kg^{-1}$ Al had lower body weights and shorter tibias than those given *ad libitum* control food, they did not differ from pair-fed controls. Thus, mallards may be more tolerant of dietary Al than some other species.

## B. Hazard Assessment

Considerable controversy remains concerning risk associated with Al exposure in birds. At water or soil pH below 5.5, Al toxicity could be expressed through the food chain because aqueous concentrations do not reach physiologically harmful levels. Plants, invertebrates, and amphibian larvae can contain high levels of Al but the bioavailability through these routes is not well known.

Insectivorous birds may be at risk in acidified environments because insects contain Al levels up to 5000 $mg \cdot kg^{-1}$ dry wt (Sadler and Lynam 1985), *Daphnia* have been reported with similar levels (Havas and Hutchinson 1983), and some estuarine invertebrates have levels exceeding 17,000 $mg \cdot kg^{-1}$ (Miles, unpublished data). Many insectivorous birds, especially those that probe or dabble in mud, may also consume substantial amounts of Al-rich soil with their foods. In acidified environments, much of this metal may be biologically active.

Predaceous aquatic birds could be intoxicated if they consumed benthic species. Amphibian tadpoles can have more than 10,000 $\mu g \cdot g^{-1}$ Al dry wt in their gut, which could be harmful if bioavailable (Sparling and Lowe 1995). Carp (*Cyprinus carpio*) may have Al levels above 3000 $\mu g \cdot g^{-1}$ (Muramoto 1981).

Herbivorous birds, including most waterfowl at certain times of the year, may have diets with high Al levels. As reported earlier, some plants under acidified conditions frequently have levels above 3000 $mg \cdot kg^{-1}$ and may be as high as 32,000 $mg \cdot kg^{-1}$ (although tea is not a waterfowl food). A diet consisting largely of these species could be toxic, but there have been no studies dealing with secondary toxicity of Al in birds.

It appears that if birds are at risk, the greatest probability of toxicity would be in rapidly growing chicks when the rate of bone deposition and P requirements are greatest. As chicks approach adult size, their tolerance appears to increase. There is little evidence that Al presents much risk to adult birds, even during egg laying, but reproduction may be impaired. Further research on the assimilation of Al and its competition with P and Ca uptake in chicks and juvenile birds is necessary to assess possible population effects.

In summary, two reviews of Al toxicity in birds reached opposing views. Scanlon (1990) concluded that "changes in water chemistry, and specifically an increase in Al availability, have been strongly implicated in reproductive impairment." Scheuhammer (1991), however, stated: "There is no convincing evidence to suggest that Pb, Cd, or Al levels in prey of most piscivorous or insectivorous wildlife dwelling in acidified environments are sufficiently high to be of general concern with respect to possible health or reproductive impairment." Scheuhammer believed that Ca deficiency in acidified waters may be a much greater problem than Al levels. We would suggest that very few species of free-ranging birds have been examined for Al toxicity and, until further studies are conducted, it is premature to make broad generalizations about the role of Al in recruitment. Although results are conflicting for the few species of birds studied under laboratory or captive conditions, data indicate considerable variation in sensitivity among species and age within a species. Therefore, Al may be potentially harmful to free-ranging birds given the right combination of diet and ambient conditions.

## VIII. Effects on Mammals

We found no information on the toxicological effects of Al on wild mammals. Studies on domesticated ungulates and laboratory mammals are available, however. Even here, acute toxicity levels are known for only a few species (Table 15).

The principal effects of elevated Al on mammals include depressed food intake, decreased weight gain in young animals or weight loss in adults, depressed serum Mg and P, and increased Al concentrations in some tissues. Earlier reviews of Al effects on mammals include those by the National Academy of Sciences (1980), Allen (1984), and Krueger et al. (1985).

### A. Ruminants

When 1450 $mg \cdot kg^{-1}$ Al (as $AlCl_3$) and 17,000 $mg \cdot kg^{-1}$ P were fed to 6-mon-old lambs (*Ovis aries*) for 76 d, food intake was depressed by 19% and weight gain declined by 45% compared to controls (Rosa et al. 1982). Elevated Al also reduced serum P, kidney P and Mg, and bone ash but increased kidney Cu. When P was raised to 42,000 $mg \cdot kg^{-1}$, many of these effects, especially depressed food intake, weight gain, and serum P, were alleviated. Effects on 10-mon-old lambs given 2000 $mg \cdot kg^{-1}$ Al were similar but included elevated residues in kidney and liver tissue (Table 16) and reduced Mn in livers; increased dietary P ameliorated the effects of dietary Al on these tissues (Valdivia et al. 1982). In yearling lambs, 2000 $mg \cdot kg^{-1}$ dietary Al with 51,000 $mg \cdot kg^{-1}$ Ca and 26,000 $mg \cdot kg^{-1}$ P increased Al absorption and retention while decreasing Mg absorption by 32% and Ca absorption by 36% (Allen et al. 1990). No effects on P or K retention were observed, but serum Ca increased and serum P decreased with elevated Al.

Cattle (*Bos taurus*) respond similarly. Dairy calves fed 2000 $mg \cdot kg^{-1}$ Al as $AlCl_3$ for 7 wk experienced 17% reduced food intake and 47% depressed weight gain (Crowe et al. 1990). Aluminum in tibias and blood serum

Table 15. Acute toxicity for mammals exposed to aluminum.

| Species | Compound | Route | $LD_{50}$ ($mg \cdot kg^{-1}$) | Reference |
|---|---|---|---|---|
| *Rattus norvegicus* | $AlCl_3$ | oral | 380 | 1 |
| *Rattus norvegicus* | $AlCl_3$ | oral | 770 | 2 |
| *Rattus norvegicus* | $Al_2(SO_4)_3$ | oral | 980 | 2 |
| *Cavia porcellus* | $AlCl_3$ | oral | 400 | 2 |
| *Lepus cuniculus* | $AlCl_3$ | oral | 400 | 2 |
| *Rattus norvegicus* | $Al(NO_3)$ | ip | 327 | 3 |
| *Mus musculus* | $Al(NO_3)$ | ip | 320 | 3 |

References: (1) Krasovskii et al. 1979; (2) Ondriecka et al. 1966; (3) Adamson et al. 1975.

Table 16. Aluminum levels ($mg \cdot kg^{-1}$ dry wt) in tissues of mammals experimentally dosed with aluminum compounds.

| Species | Tissue | Treated | Control | Comments | References |
|---|---|---|---|---|---|
| *Ovies aries*[a] | liver | 21.8 | 8.6 | fed 2000 $mg \cdot kg^{-1}$ as Al-citrate | 1 |
| | kidney | 23.0 | 10.5 | as above | |
| | muscle | 14.6 | 8.3 | as above | |
| | brain | 21.7 | 16.0 | as above | |
| *Ovies aries*[a] | liver | 21.1 | 9.1 | fed 2000 $mg \cdot kg^{-1}$ as $AlCl_3$ | 2 |
| | kidney | 21.8 | 12.9 | as above | |
| | muscle | 14.3 | 9.3 | as above | |
| | brain | 19.3 | 19.3 | as above | |
| *Bos taurus*[b] | liver | 36.4 | 24.7 | fed 1200 $mg \cdot kg^{-1}$ as $AlCl_3$ | 3 |
| | kidney | 24.0 | 20.0 | as above | |
| | muscle | 16.8 | 13.6 | as above | |
| | brain | 33.1 | 27.5 | as above | |
| Dairy calves[b] | tibia shaft | 21.2 | 7.9 | fed 2000 $mg \cdot kg^{-1}$ as $AlCl_3$ | 4 |
| | tibia joint | 35.0 | 25.6 | as above | |
| *Lepus cuniculus* | liver | 17.0 | 1.3 | 10.8 mg sc weekly/4 wk | 5 |
| | kidney | 21.0 | 8.6 | as above | |
| | muscle | 0.3 | 1.0 | as above | |
| | brain | 2.0 | 0.8 | as above | |
| | bone | 92.0 | 21.0 | as above | |
| *Lepus cuniculus* | hair | 532 | 98 | as above | 6 |
| *Lepus cuniculus*[c] | liver | 40.6 | 9.6 | 21.6 mg sc, 20 injections | 7 |
| | kidney | 238 | 5.4 | as above | |
| | muscle | 1.2 | 0.9 | as above | |
| | brain | 6.3 | 5.8 | as above | |
| | bone | 79.4 | 31.4 | as above | |
| *Lepus cuniculus*[d] | liver | 246 | 2.2 | as above | 7 |
| | kidney | 1290 | 4.2 | as above | |
| | muscle | 2.5 | 1.8 | as above | |
| | brain | 6.8 | 3.2 | as above | |
| | bone | 90.3 | 18.7 | as above | |

*(continued)*

Table 16. (*Continued*)

| Species | Tissue | Treated | Control | Comments | References |
|---|---|---|---|---|---|
| *Rattus norvegicus* | liver | 29.9 | 15.5 | fed 2835 $mg \cdot kg^{-1}$ for 8 d | 8 |
| | brain | 10.8 | 7.1 | as above | |
| | kidney | 9.0 | 8.6 | as above | |
| | blood | 10.8 | 6.5 | as above | |
| | femur | 912 | 702 | as above | |
| *Rattus norvegicus*[e] | tibia | 1349 | 971 | fed 1106 $mg \cdot kg^{-1}$ for 30 d, no citrate | 9 |
| | kidney | 351 | 297 | as above | |
| | tibia | 1861 | 971 | as above with 5.2 $\mu$mol citrate | |
| | kidney | 459 | 297 | as above | |

[a]Concentration converted from wet weight to dry weight using dry matter content of tissues (%) as liver, 30.77; kidney, 22.46; muscle, 28.01; brain, 23.25 (Valdivia et al. 1982).

[b]Expressed as ash weight.

[c]Lactating does.

[d]Nonlactating does.

[e]Weanling rats, expressed as wet weight.

References: (1) Valdivia et al. 1982; (2) Valdivia-Rodriguez 1977; (3) Valdivia et al. 1978; (4) Crowe et al. 1990; (5) Yokel 1987; (6) Yokel 1982; (7) Yokel 1984; (8) Ondriecka et al. 1966; (9) Ecelbarger and Greger 1991.

increased compared to control animals (Table 16) and some enzymes (e.g., alkaline phosphatase) also increased. In lactating cows (Allen 1984), 2000 $mg \cdot kg^{-1}$ Al as aluminum citrate depressed food intake and reduced serum Mg by 25% and serum P by 34% over a 56-d feeding trial.

The apparent depressant effect of dietary Al on Mg absorption and serum levels in ruminants has led researchers (Allen and Robinson 1980) to associate the metal in forage with grass tetany. Clinical signs of this disease include bellowing in cattle, muscular fasiculations or twitches, galloping in an undirected fashion, convulsions, hypersensitivity, hyperurination, and death. The cause of the disease seems to be hypomagnesemia, and it appears more frequently in heavily lactating females than in males or nonlactating females. Allen and Robinson (1980) identified 11 incidents of the disease in Louisiana and Tennessee over a 28-mon period. Forage types included ryegrass, ryegrass mixed with grain, or fescue. Aluminum levels in forage ranged from 2000 to 14,500 $mg \cdot kg^{-1}$ dry wt, and Mg levels were typically less than 2000 $mg \cdot kg^{-1}$. Animals that died from tetany had 1630–3390 $mg \cdot kg^{-1}$ Al in their rumens, whereas nontetanus animals had 330–510 $mg \cdot kg^{-1}$. Rumen Mg levels did not differ between the two groups.

The greatest source of Al for grazers is probably soil contamination of forage. Robinson et al. (1984) found that ryegrass forage generally had Al levels less than 100 $mg \cdot kg^{-1}$ in spring, but when the condition of pastures deteriorated in autumn, concentrations exceeded 1000 $mg \cdot kg^{-1}$ and rumen levels reached 2200 $mg \cdot kg^{-1}$: most of this was attributed to soil contamination. Aluminum can also be consumed with drinking water, especially in areas where pH of water is 5.0 or less (Allen 1984), but this is probably not an important source of exposure.

### B. Laboratory Animals

Repeated injections of 0, 10.8, and 21.6 $mg \cdot kg^{-1}$ Al body weight over 25 d produced sublethal effects in lactating rabbits (*Lepus cuniculus*) (Yokel 1984). The most obvious differences included weight loss by does and reduced weight gain by their young. All of the does given 21.6 $mg \cdot kg^{-1}$ developed abnormal posture, splayed limbs, and impaired locomotion. Milk from does on the high dosage contained an average of 6 $mg \cdot kg^{-1}$ Al but did not result in any physiological problems in young. When 10.8 $mg \cdot kg^{-1}$ was repeatedly injected into pregnant does, 58% of their fetuses were stillborn or dead within 2 d after the injections. Surviving young appeared to have problems in learning and memory. Neonates given 10.8 $mg \cdot kg^{-1}$ weekly for 4 wk decreased milk consumption by 20%, and had lower weight gain and decreased calcification of ulnas compared to those injected with a placebo (Yokel 1987). However, no differences in survivorship or in behavioral tests were noted between control and treatment groups. Although wild animals would not be exposed subcutaneously to Al, Yokel's work is instructive in demonstrating that the metal can be passed *trans utero* and via lactation, and that fetuses are the most vulnerable life stage in rabbits.

One of the earliest studies demonstrating Al toxicity in mammals was conducted by Ondriecka et al. (1966), who determined the $LD_{50}$ for $AlCl_3$ in rats (*Rattus norvegicus*) to be 770 $mg \cdot kg^{-1}$ Al and 980 $mg \cdot kg^{-1}$ for the sulfate form. Also, chronic exposure of rats to 19.3 $mg \cdot kg^{-1} \cdot d^{-1}$ for 180–390 d had no apparent effect on adult physiology but did reduce growth rates in young animals starting with the second litter. When they exposed mice (*Mus musculus*) to dietary Al at 170 and 355 $mg \cdot kg^{-1}$, Ondriecka et al. (1966) found that food consumption was reduced but not to the point of decreasing growth. Phosphorus retention was significantly depressed, and even a single oral dose of 188 $mg \cdot kg^{-1}$ Al reduced $^{32}P$ uptake by 92% in the blood and 96% in femurs. Long-term exposure (>50 d) to 365 $mg \cdot kg^{-1}$ decreased ATP and increased ADP in mice. Several other studies have examined toxicity in laboratory rodents (Krueger et al. 1985; Scheuhammer 1991). In general, dietary levels greater than 100 $mg \cdot kg^{-1}$ may produce physiological effects; lethal acute doses are approximately 1 $g \cdot kg^{-1}$; and principal organs of accumulation include bones, brain, and kidneys.

### C. Hazard Assessment

In general, mammals appear to be at moderate to minimal risk of Al toxicity. Aluminum is not accumulated at sufficient levels in soft animal tissues to produce secondary toxicity in carnivores. Cattle may be prone to grass tetany when Mg levels in soils and forage are low, but only a few incidents of the disease have been reported. Unless the soil is acidic, Al ingested with soil is probably insoluble and has low biological availability; gut pH and retention time would be important in determining how much of this metal would be assimilated. Soluble Al in acidic forest soils may pose a risk, but problems would largely be manifested as reduced feeding and weight gain. Unfortunately, there have been no studies dealing with wildlife, so no generalizations can be made about Al toxicity in these species.

## IX. Remediation in Natural Systems

Remediation of Al toxicity in natural environments seems to be largely a matter of regulating pH. Because Al solubility is very low when pH is greater than 5.5, use of buffers on forest soils, surface waters, and watersheds can be effective in reducing the more biologically active, inorganic, monomeric Al species. The most common method of buffering soils and waters in North America and Europe is application of powdered limestone ($CaCO_3$) to watersheds and water. A variety of application techniques including aerial application, dispersal onto ice cover or water surfaces, and periodic mixing with flowing waters have been developed. Reviews of techniques, criteria for determining dosages and application methods, and biological and chemical effects are given by Olem (1991).

Most studies on the effects of liming on Al concentration and speciation have occurred in freshwater environments. Marcus (1988) determined that the average concentration of Al in 20 lakes surveyed remained at approximately 40 $\mu g \cdot L^{-1}$ both pre- and posttreatment with lime. However, the species of the metal examined or the duration between time of application and posttreatment chemistry analysis were not mentioned. In theory, as pH increases from 4.0 to 5.5, concentrations of monomeric Al should decrease more rapidly than total Al because of decreasing Al solubility within this range. Although pH and Ca may increase within a few hours after liming, Al concentrations (both total and monomeric) decrease over a period of weeks.

Liming was only partially successful in reducing Al levels among nine lakes in the Adirondack Mountains of New York (Gloss et al. 1989). Six of the nine lakes demonstrated decreases of monomeric Al from 1% to 47% of the pretreatment levels. The other lakes demonstrated increases of monomeric Al as high as 240% of the original values. In comparison, only four lakes showed declines in total metal levels. Acidity in all nine lakes declined temporarily by 0.3–1.5 pH units. Reductions in Al concentrations and maintenance of elevated pH were compromised by continual inputs of acidic water.

Liming has been a standard practice in Sweden for many years (Hasselrot and Hultberg 1984). Typically, 50–70 $kg \cdot ha^{-1}$ of limestone is applied to the watershed for each year of desired remediation. If applied directly to water or ice cover, 10–30 $mg \cdot L^{-1}$ of water volume is used. Duration of buffering varies with renewal times of lakes and stream flow rates, but the goal is often 5 yr. Elevation of pH and Ca and reduction of Al are the most common and obvious changes following liming. Total Al may drop from 300–700 $\mu g \cdot L^{-1}$ before treatment of lakes to 50–100 $\mu g \cdot L^{-1}$ after liming. In streams, the concentrations may decrease from 700–1200 $\mu g \cdot L^{-1}$ to 150–500 $\mu g \cdot L^{-1}$. The decreases in total Al probably result from precipitation. Levels of Al dropped from approximately 300 $\mu g \cdot L^{-1}$ to 120 $\mu g \cdot L^{-1}$ shortly after liming Lake Hovvatn, Norway, and remained depressed for 1.5–2 yr (Wright 1985). Coincidentally, pH increased from 4.5 to 7.4 in one treated catchment of the lake and from 4.5 to 6.4 in another.

Driscoll et al. (1989) determined that inorganic monomeric Al dropped more than 70% from 202 $\mu g \cdot L^{-1}$ to 54 $\mu g \cdot L^{-1}$ after 4 wk when two lakes in the Adirondack Mountains were treated with limestone at approximately 1 metric ton$\cdot ha^{-1}$. Other chemical changes associated with liming included increases in dissolved inorganic carbon concentrations and in acid-neutralizing capacity. Waters above the thermocline were most notably affected by liming in the spring, but there was some buffering in deeper waters after fall turnover. Both ponds had rapid lake renewal times of less than 174 d with continuous inflows of acidified water, and one pond became reacidified after 7 mon. When the pH of this lake fell below 5.5, monomeric Al increased dramatically. The other pond continued to show remediation of pH and Al after 14 mon.

As stated previously, it is possible that liming can temporarily increase Al and other metal toxicity to fish. As pH increases above 4.5, concentrations of colloidal Al may also increase. In the laboratory, colloidal Al was found to adhere to gill filaments of fish and appeared to be more toxic than dissolved, monomeric Al (Muniz and Leivestad 1980). The respiration rate of three species of mayflies increased within the pH range of 5.5–6.5 when 500 $\mu g \cdot L^{-1}$ was added, probably because of Al adhering onto gill filaments (Herrmann and Andersson 1986). It is possible that a rapid elevation of pH from less than 5.0 to more than 5.5 could increase adsorption of Al by insect larvae.

Under some conditions, lakes may be able to recuperate from acidic, elevated Al conditions without remediation when influx of acid-producing ions is curtailed. The Sudbury region of Ontario, Canada, is an extensively studied region surrounding a metal smelter. For more than 30 yr, atmospheric deposition of $SO_4^{2-}$ and metals severely decreased water quality in many surrounding lakes and streams (Keller and Yan 1991). Since 1980, rates of $SO_4^{2-}$ deposition have decreased substantially, and there is considerable variation in water quality among neighboring lakes, depending on proximity and direction from the smelter. Total Al in acidic lakes within the region has dropped by 33%–84% since emissions were reduced. Acidity has also declined, although the pH of many lakes remains below 5.5. Biological recovery has been most obvious in lakes that have undergone the greatest improvements in water quality.

Aluminum toxicity also may be ameliorated by adding certain cations such as $Ca^{2+}$, $Mg^{2+}$, $Na^{+}$, $K^{+}$, or Sr to soil. Parker et al. (1989) determined that in wheat the addition of $Ca^{2+}$ was approximately 1.5 times more efficient (wt/wt) in ameliorating Al phytotoxicity than $Mg^{2+}$ and approximately 15 times more efficient than $Na^{+}$. Gypsum ($CaSO_4$) would be an inexpensive material to ameliorate Al toxicity in acid soils. Adding P can also reduce toxic effects of Al by compensating for reduced P.

In summary, liming watersheds and surface waters is effective in reducing Al levels and ameliorating its toxicity. Differences in the degree of success are related to the effectiveness of reducing acidity. Success may be more difficult to achieve in lakes with continual inputs of acidic water with elevated Al or in streams. Waters with low renewal rates may show appreciable long-term (to 5 yr) improvement in concentrations of Al, especially the more toxic monomeric species. Less is known about the effects of liming soils and reducing toxicity in terrestrial habitats, although limestone is a standard method of reducing soil acidity, and similar results should be expected.

## Summary

Aluminum is extremely common throughout the world and is innocuous under circumneutral or alkaline conditions. However, in acidic environments, it can be a major limiting factor to many plants and aquatic organisms. The greatest concern for toxicity in North America occurs in areas that are affected by wet and dry acid deposition, such as eastern Canada and the northeastern U.S. Acid mine drainage, logging, and water treatment plant effluents containing alum can be other major sources of Al. In solution, the metal can combine with several different agents to affect toxicity. In general, Al hydroxides and monomeric Al are the most toxic forms. Dissolved organic carbons, F, $PO_3^{3-}$ and $SO_4^{2-}$ ameliorate toxicity by reducing bioavailability.

Elevated metal levels in water and soil can cause serious problems for some plants. Algae tend to be both acid- and Al tolerant and, although some species may disappear with reduced pH, overall algae productivity and biomass are seldom affected if pH is above 3.0. Aluminum and acid toxicity tend to be additive to some algae when pH is less than 4.5. Because the metal binds with inorganic P, it may reduce P availability and reduce productivity. Forest die-backs in North America involving red spruce, Fraser fir, balsam fir, loblolly pine, slash pine, and sugar maples have been ascribed to Al toxicity, and extensive areas of European forests have died because of the combination of high soil Al and low pH. Extensive research on crops has produced Al-resistant cultivars and considerable knowledge about mechanisms of and defenses against toxicity. Very low Al levels may benefit some plants, although the metal is not recognized as an essential nutrient. Hyperaccumulator species of plants may concentrate Al to levels that are toxic to herbivores.

Toxicity in aquatic invertebrates is also acid dependent. Taxa such as Ephemeroptera, Plecoptera, and Cladocera are sensitive and may perish when Al is less than 1 $mg \cdot L^{-1}$ whereas dipterans, molluscs, and isopods seem to be tolerant. In Al-sensitive species, elevated levels ($\sim 500\ \mu g \cdot L^{-1}$) affect ion regulation and respiratory efficiency. Toxicity tends to be greatest near a species' threshold of pH sensitivity. At lower pHs, Al may have a slight ameliorative effect by interfering with $H^+$ transport across membranes. Aquatic invertebrates can accumulate very high levels of Al, but most of this appears to be through adsorption rather than assimilation. Aluminum concentrations may be as high as 5000 $mg \cdot kg^{-1}$ in insects and greater than 17,000 $mg \cdot kg^{-1}$ in other invertebrates.

The bulk of the research on Al toxicity in animals has been on fish, with more than 200 papers published since 1980. Toxicity in many fish species is bimodal in that it causes asphyxiation at higher pHs ($\sim$5.5–6.5) and interferes with electrolyte balance at lower pH. Sensitivity varies considerably with age and among species. For example, eggs and larval fish tend to be more sensitive to Al and pH than older fish. Dissolved organic carbon and F appear to ameliorate the toxic effects of Al at low pH. Because of bimodal toxicity with respect to pH, attempts to reduce acidity through liming or by mixing low- and high-pH waters may actually increase fish mortality at the point of mixing. The importance of Al in limiting fish distributions relative to other factors such as pH, divalent cations, and DOC is not clearly understood.

Aluminum and pH interact in much the same way on amphibians. At pH above 5.0, tadpole $LC_{50}$s for Al range from 400 to 750 $\mu g \cdot L^{-1}$. Newly hatched tadpoles are among the most sensitive life stages, followed by embryos and older tadpoles. As with fish, Al may temporarily ameliorate $H^+$ toxicity, but this effect probably has no consequence on the overall reproductive output of amphibians. The consequences of Al toxicity are, in part, affected by habitat in that species that breed in temporary ponds may be at greater risk than those breeding in larger ponds and lakes. Limited data on body burdens in tadpoles indicate that they may contain very high levels and could be toxic to predators.

The possible effects of toxicity in birds are controversial. If the metal is important, it probably acts through the food chain by reducing overall abundance or composition of invertebrate foods. Some food organisms may contain sufficient Al to pose a hazard to avian predators. Dietary studies of Al have been contradictory in that some have shown no effects on reproduction, growth, or survival, whereas others have demonstrated very significant alterations of all three factors. There is substantial interspecific variation in response to Al, and young, rapidly growing birds are much more sensitive to dietary Al than older birds. In Al-sensitive species, the metal binds with P and results in rickets-like symptoms. The effects of dietary Al are augmented by low dietary P and Ca.

Essentially nothing is known about the toxic effects of Al on wild mammals and little about free-ranging domestic species. Theoretically, mammals may ingest sufficient Al in soil to cause problems such as grass tetany, but observations of such problems in wild ungulates have not been published. As in other species, the metal reduces P availability and may bind with Mg to make it less available. Laboratory studies have shown that Al can be transferred to embryos and through milk to nursing young.

The essence of remediating Al is elevating pH of water or soil. One of the most effective means of raising pH in natural environments is broadcasting lime. The addition of divalent cations to soil can also be effective. If the source of anthropogenic acidification can be removed or reduced, pH may rise through natural processes.

## Acknowledgments

We thank K. Miles, K. Schreiber, and T. Haines for their excellent recommendations on improving earlier drafts of this manuscript. A. Kahn provided editorial support.

## References

Adams MB, Eagar C (1992) Impacts of acidic deposition on high-elevation spruce-fir forests: results from the Spruce-Fir Research Cooperative. For Ecol Manage 51:195–205.

Adamson RH, Canellos GP, Sieber SM (1975) Studies on the antitumor activity of gallium nitrate (NSC-15200) and other group IIIa metal salts. Cancer Chemother Rep 59:599–610.

Albers PH, Camardese MB (1993) Effects of acidification on metal accumulation by aquatic plants and invertebrates. 2. Wetlands, ponds and small lakes. Environ Toxicol Chem 12:969–976.

Allen IRH, Ritter JA (1977) Salmonid terminology. J Cons Int Explor Mer 37:293–299.

Allen VG, Robinson DL (1980) Occurrence of Al and Mn in grass tetany cases and their effects on the solubility of Ca and Mg in vitro. Agron J 72:957–960.

Allen VG (1984) Influence of dietary aluminum on nutrient utilization in ruminants. J Anim Sci 59:836–844.

Allen VG, Fontenot JP, Rahnema SH (1990) Influence of aluminum citrate and citric acid on mineral metabolism in wether sheep. J Anim Sci 68:2496–2505.

American Fisheries Society (1980) A list of common and scientific names of fishes from the United States and Canada, 4th Ed. Spec Publ 12, American Fisheries Society, Bethesda, MD.

Baes CF III, McLaughlin SB (1987) Trace metal uptake and accumulation in trees as affected by environmental pollution. In: Hutchinson TC, Meems KM (eds) Effects of Atmospheric Pollutants on Forests, Wetlands, and Agricultural Ecosystems. Vol. G16, NATO ASI Series. Springer-Verlag, Berlin, pp 307–319.

Baker JP, Schofield CL (1982) Aluminum toxicity to fish in acidic waters. Water Air Soil Pollut 18:289–309.

Baker JP, Christensen SW (1991) Effects of acidification on biological communities in aquatic ecosystems. In: Charles DF (ed) Acidic Deposition and Aquatic Ecosystems. Regional Case Studies. Springer-Verlag, New York, pp 83–106.

Battram JC (1988) The effects of aluminium and low pH on chloride fluxes in the brown trout, *Salmo trutta* L. J Fish Biol 32:937–947.

Bendell-Young LI, Harvey HH (1986) Metal concentration and calcification of bone of white sucker (*Catostomus commersoni*) in relation to lake pH. Water Air Soil Pollut 30:657–664.

Bengtsson B, Asp H, Jensen P, Berggren D (1988) Influence of aluminium on phosphate and calcium uptake in beech (*Fagus sylvatica*) grown in nutrient solution and soil solution. Physiol Plant 74:299–305.

Berg DJ, Burns TA (1985) The distribution of aluminum in the tissues of three fish species. J Freshwater Ecol 3:113–120.

Biesinger KE, Christensen GM (1972) Effects of various metals on survival, growth, reproduction, and metabolism of *Daphnia magna*. J Fish Res Board Can 29:1691–1700.

Birchall JD, Exley C, Chappell JS, Phillips MJ (1989) Acute toxicity of aluminium to fish eliminated in silicon-rich acid waters. Nature 338:146–148.

Blamey FPC, Edmeades DC, Wheeler DM (1990a) Role of root cation-exchange capacity in differential aluminum tolerance of *Lotus* species. J Plant Nutr 13:729–744.

Blamey FPC, Wheeler DM, Edmeades DC, Christie RA (1990b) Independence of differential aluminum tolerance in *Lotus* on changes in rhizosphere pH or excretion of organic ligands. J Plant Nutr 13:713–728.

Booth CE, McDonald DG, Simons BP, Wood CM (1988) Effects of aluminum and low pH on net ion fluxes and ion balance in the brook trout (*Salvelinus fontinalis*). Can J Fish Aquat Sci 45:1563–1574.

Bortolotti GR, Barlow JC (1988) Some sources of variation in the elemental composition of bald eagle feathers. Can J Zool 66:1948–1951.

Bortolotti GR, Szuba KJ, Naylor BJ, Bendell JF (1988) Stability of mineral profiles of spruce grouse feathers. J Wildl Manage 52:736–743.

Brouard D, Van Coillie R, Vigneault Y, Thellen C (1989) Bioaccumulation et effets sous-letaux de l'aluminium chez des oeufs et juveniles de salmonides: experimentation en laboratoire et en milieu naturel. Can Tech Rep Fish Aquat Sci 1714:9–31.

Brown DJA (1981) The effects of various cations on the survival of brown trout, *Salmo trutta* at low pHs. J Fish Biol 18:31–40.

Brown DJA (1983) Effect of calcium and aluminum concentrations on the survival of brown trout (*Salmo trutta*) at low pH. Bull Environ Contam Toxicol 30:582–587.

Brumbaugh WG, Kane DA (1985) Variability of aluminum concentrations in organs and whole bodies of smallmouth bass (*Micropterus dolomieui*). Environ Sci Technol 19:828–831.

Buckler DR, Mehrle PM, Cleveland L, Dwyer FJ (1987) Influence of pH on the toxicity of aluminum and other inorganic contaminants to east coast striped bass. Water Air Soil Pollut 35:97–106.

Buergel PM, Soltero RA (1983) The distribution and accumulation of aluminum in rainbow trout following a whole-lake alum treatment. J Freshwater Ecol 2:37–44.

Burton TM, Allan JW (1986) Influence of pH, aluminum, and organic matter on stream invertebrates. Can J Fish Aquat Sci 43:1285–1289.

Capdevielle MC, Scanes C (1995a) Effect of dietary acid and aluminum on growth and growth related hormones in young chickens. Toxicol Appl Pharmacol 133: 164–171.

Capdevielle MC, Scanes C (1995b) Effect of dietary acid or aluminum on growth and growth related hormones in young mallard ducklings (*Anas platyrhynchos*). Arch Environ Contam Toxicol (in press).

Carriere D, Fischer KL, Peakall DB, Anghern P (1986) Effects of dietary aluminum sulphate on reproductive success and growth of ringed turtle-doves (*Streptopelia risoria*). Can J Zool 64:1500–1505.

Catling PM, Freedman B, Stewart C, Kerekes JJ, Lefkovitch LP (1986) Aquatic plants of acid lakes in Kejimkujik National Park, Nova Scotia; floristic composition and relation to water chemistry. Can J Bot 64:724–729.

Chevalier G, Gauthier L, Moreau G (1985) Histopathological and electron microscopic studies of gills of brook trout, *Salvelinus fontinalis*, from acidified lakes. Can J Zool 63:2062–2070.

Clark KL, Hall RJ (1985) Effects of elevated hydrogen ion and aluminum concentrations on the survival of amphibian embryos and larvae. Can J Zool 63:116–123.

Clark KL, LaZerte BD (1985) A laboratory study of the effects of aluminum and pH on amphibian eggs and tadpoles. Can J Fish Aquat Sci 42:1544–1551.

Cleveland L, Little EE, Hamilton SJ, Buckler DR, Hunn JB (1986) Interactive toxicity of aluminum and acidity to early life stages of brook trout. Trans Am Fish Soc 115:610–620.

Cleveland L, Little EE, Wiedmeyer RH, Buckler DR (1989) Chronic no-observed-effect concentrations of aluminum for brook trout exposed in low-calcium, dilute acidic water. In: Lewis TE (ed) Environmental Chemistry and Toxicology of Aluminum. Lewis Publishers, Chelsea, MI, pp 229–245.

Cleveland L, Buckler DR, Brumbaugh WG (1991a) Residue dynamics and effects of aluminum on growth and mortality in brook trout. Environ Toxicol Chem 10: 243–248.

Cleveland L, Little EE, Ingersoll CG, Wiedmeyer RH, Hunn JB (1991b) Sensitivity of brook trout to low pH, low calcium and elevated aluminum concentrations during laboratory pulse exposures. Aquat Toxicol 19:303–318.

Cook RB, Jager HI (1991) Upper Midwest. In: Charles DF (ed) Acidic Deposition and Aquatic Ecosystems. Springer-Verlag, New York, pp 421–466.

Correa M, Coler RA, Yin C (1985) Changes in oxygen consumption and nitrogen metabolism in the dragonfly *Somatochlora cingulata* exposed to aluminum in acid waters. Hydrobiologia 121:151–156.

Cronan CS, Schofield CL (1990) Relationships between aqueous aluminum and acidic deposition in forested watersheds of North America and northern Europe. Environ Sci Technol 24:1100–1105.

Crowder A (1991) Acidification, metals and macrophytes. Environ Pollut 71:171–203.

Crowe NA, Neathery MW, Miller WJ, Muse LA, Crowe CT, Varnadoe JL, Blackmon DM (1990) Influence of high dietary aluminum on performance and phosphorus bioavailability in dairy calves. J Dairy Sci 73:808–818.

Cummins CP (1986) Effects of aluminium and low pH on growth and development in *Rana temporaria* tadpoles. Oecologia 69:248–252.

Dalziel TRK, Morris R, Brown DJA (1986) The effects of low pH, low calcium concentrations and elevated aluminium concentrations on sodium fluxes in brown trout, *Salmo trutta* L. Water Air Soil Pollut 30:569–577.

Dietrich D, Schlatter C (1989a) Low levels of aluminum causing death of brown trout (*Salmo trutta fario* L.) in a Swiss alpine lake. Aquat Sci 51:279–295.

Dietrich D, Schlatter C (1989b) Aluminium toxicity to rainbow trout at low pH. Aquat Toxicol 15:197–212.

Dietrich D, Schlatter CH, Blau N, Fischer M (1989) Aluminium and acid rain: mitigating effects of NaCl on aluminium toxicity to brown trout (*Salmo trutta fario*) in acid water. Toxicol Environ Chem 19:17–23.

Driscoll CT Jr, Baker JP, Bisogni JJ Jr, Schofield CL (1980) Effect of aluminium speciation on fish in dilute acidified waters. Nature 284:161–164.

Driscoll CT, Schecher WD (1988) Aluminum in the environment. In: Sigel H, Sigel A (eds) Metal Ions in Biological Systems, Vol. 24. Aluminum and its role in Biology. Marcel Dekker, New York, pp 59–122.

Driscoll CT, Ayling WA, Fordham GF, Oliver LM (1989) Chemical response of lakes treated with $CaCO_3$ to reacidification. Can J Fish Aquat Sci 46:258–267.

Driscoll CT, Schecher WD (1990) The chemistry of aluminum in the environment. Environ Geochem Health 12:28–50.

Driscoll CT, Newton RM, Gubala CP, Baker JP, Christensen SW (1991) Adirondack Mountains. In: Charles DF (ed) Acidic Deposition and Aquatic Ecosystems. Springer-Verlag, New York, pp 133–202.

Ecelbarger CA, Greger JL (1991) Dietary citrate and kidney function affect aluminum, zinc and iron utilization in rats. J Nutr 121:1755–1762.

Eddy FB, Talbot C (1985) Sodium balance in eggs and dechorionated embryos of the Atlantic salmon *Salmo salar* L. exposed to zinc, aluminium and acid waters. Comp Biochem Physiol 81C:259–266.

Eldhuset T, Goransson A, Ingestad T (1987) Aluminum toxicity in forest tree seedlings. In: Hutchinson TC, Meems KM (eds) Effects of Atmospheric Pollutants on Forests, Wetlands and Agricultural Ecosystems. Vol. G16, NATO ASI Series. Springer-Verlag, Berlin, pp 401–409.

Eriksson MOG, Henrikson L, Oscarson HG (1989) Metal contents in liver tissues of non-fledged goldeneye, *Bucephala clangula*, ducklings: a comparison between samples from acidic, circumneutral, and limed lakes in south Sweden. Arch Environ Contam Toxicol 18:255–260.

Fageria NK, Baligar VC, Wright RJ (1988) Aluminum toxicity in crop plants. J Plant Nutr 11:303–319.

Firling CE, Severson AR, Hill TA (1994) Aluminum effects on blood chemistry and long bone development in the chick embryo. Arch Toxicol 68:541–547.

Fivelstad S, Leivestad H (1984) Aluminium toxicity to Atlantic salmon (*Salmo salar* L.) and brown trout (*Salmo trutta* L.): Mortality and physiological response. Vol. 61. National Swedish Board of Fisheries, Institute of Freshwater Research, Drottningholm, Sweden, pp 69–77.

Folkeson L, Nyholm NEI, Tyler G (1990) Influence of acidity and other soil properties on metal concentrations in forest plants and animals. Sci Total Environ 96: 211–233.

Foy CD (1988) Plant adaptation to acid, aluminum-toxic soils. Commun Soil Sci Plant Anal 19:959–987.

Freda J (1986) The influence of acidic pond water on amphibians: a review. Water Air Soil Pollut 30:439–450.

Freda J, Cavdek V, McDonald DG (1989) Role of organic complexation in the toxicity of aluminum to *Rana pipiens* embryos and *Bufo americanus* tadpoles. Can J Fish Aquat Sci 47:217–224.

Freda J, McDonald DG (1990) Effects of aluminum on the leopard frog, *Rana pipiens*: life stage comparisons and aluminum uptake. Can J Fish Aquat Sci 47: 210–216.

Freda J (1991) The effects of aluminum and other metals on amphibians. Environ Pollut 71:305–328.

Freedman BC, Hutchinson TC (1986) Aluminum in terrestrial ecosystems. In: Havas M, Jaworski JF (eds) Aluminum in the Canadian Environment. NRCC 24759. National Research Council of Canada, Ottawa, pp 129–152.

Frick KG, Herrmann J (1990) Aluminum accumulation in a lotic mayfly at low pH—a laboratory study. Ecotoxicol Environ Saf 19:81–88.

Gagen CJ, Sharpe WE (1987) Net sodium loss and mortality of three salmonid species exposed to a stream acidified by atmospheric deposition. Bull Environ Contam Toxicol 39:7–14.

Gilani SH, Chatzinoff M (1981) Aluminum poisoning and chick embryogenesis. Environ Res 24:1–5.

Glooschenko V, Blancher P, Herskowitz J, Fulthorpe R, Rang S (1986) Association of wetland acidity with reproductive parameters and insect prey of the eastern kingbird (*Tyrannus tyrannus*) near Sudbury, Ontario. Water Air Soil Pollut 30: 553–567.

Gloss SP, Schofield CL, Marcus MD (1989) Liming and fisheries management guidelines for acidified lakes in the Adirondack region. US Fish Wildl Serv Biol Rep 80(40.27).

Goldschmidt VM (1958) Geochemistry. Oxford University Press, London.

Goss GG, Wood CM (1988) The effects of acid and acid/aluminum exposure on circulating plasma cortisol levels and other blood parameters in the rainbow trout, *Salmo gairdneri*. J Fish Biol 32:63–76.

Gostomski F (1990) The toxicity of aluminum to aquatic species in the US. Environ Geochem Health 12:51–54.

Gunn JM, Noakes DLG (1987) Latent effects of pulse exposure to aluminum and low pH on size, ionic composition, and feeding efficiency of lake trout (*Salvelinus namaycush*) alevins. Can J Fish Aquat Sci 44:1418–1424.

Haines TA (1981) Acidic precipitation and its consequences for aquatic ecosystems: a review. Trans Am Fish Soc 110:669–707.

Haines TA, Pauwels SJ, Jagoe CH, Norton SA (1987) Effects of acidity-related water and sediment chemistry variables on trace metal burdens in brook trout (*Salvelinus fontinalis*). Ann Soc R Zool Belg 117(suppl 1):45–55.

Hall LW Jr, Pinkney AE, Horseman LO, Finger SE (1985) Mortality of striped bass larvae in relation to contaminants and water quality in a Chesapeake Bay tributary. Trans Am Fish Soc 114:861–868.

Hamilton SJ, Haines TA (1989) Bone characteristics and metal concentrations in white suckers (*Catostomus commersoni*) from one neutral and three acidified lakes in Maine. Can J Fish Aquat Sci 46:440–446.

Haridasan GE (1982) Aluminum accumulation by some certain cerrado species of central Brasil. Plant Soil 65:265–273.

Harvey HH, McArdle JM (1986) Physiological responses of rainbow trout *Salmo gairdneri* exposed to Plastic Lake inlet and outlet stream waters. Water Air Soil Pollut 30:687–694.

Harvey HH, Whelpdale DM (1986) On the prediction of acid precipitation events and their effects on fishes. Water Air Soil Pollut 30:579–586.

Hasselrot B, Hultberg H (1984) Liming of acidified Swedish lakes and streams and its consequences for aquatic ecosystems. Fisheries (Bethesda) 9:4–9.

Havas M (1985) Aluminum bioaccumulation and toxicity to *Daphnia magna* in soft water at low pH. Can J Fish Aquat Sci 42:1741–1748.

Havas M (1986a) Aluminum chemistry of inland waters. In: Havas M, Jaworski JF (eds) Aluminum in the Canadian Environment. NRCC 24759. National Research Council of Canada, Ottawa, pp 51–77.

Havas M (1986b) Effects of aluminum on aquatic biota. In: Havas M, Jaworski JF (eds) Aluminum in the Canadian Environment. NRCC 24759. National Research Council of Canada, Ottawa, pp 79–127.

Havas M, Hutchinson TC (1982) Aquatic invertebrates from the Smoking Hills, N.W.T.: Effects of pH and metals on mortality. Can J Fish Aquat Sci 39:890–903.

Havas M, Hutchinson TC (1983) Effect of low pH on the chemical composition of aquatic invertebrates from tundra ponds at the Smoking Hills, N.W.T., Canada. Can J Zool 61:241–249.

Havas M, Likens GE (1985) Toxicity of aluminum and hydrogen ions to *Daphnia catawba, Holopedium gibberum, Chaoborus punctipennis,* and *Chironomus anthrocinus* from Mirror Lake, New Hampshire. Can J Zool 63:1114–1119.

Havens KE, DeCosta J (1987) The role of aluminium contamination in determining phytoplankton and zooplankton responses to acidification. Water Air Soil Pollut 33:277–293.

Havens KE, Heath RT (1989) Acid and aluminum effects on freshwater zooplankton: an *in situ* mesocosm study. Environ Pollut 62:195–211.

Havens KE, Heath RT (1990) Phytoplankton succession during acidification with and without increasing aluminum levels. Environ Pollut 68:129–145.

Haynes RJ, Klimstra WD (1975) Some properties of coal spoilbank and refuse materials resulting from surface-mining coal in Illinois. Ill Inst Environ Qual Doc. No. 75-21, Springfield, IL.

Heming TA, Blumhagen KA (1988) Plasma acid-base and electrolyte states of rainbow trout exposed to alum (aluminum sulfate) in acidic and alkaline environments. Aquat Toxicol 12:125–140.

Henny CJ, Blus LJ, Grove RA, Thompson SP (1991) Accumulation of trace elements and organochlorines by surf scoters wintering in the Pacific Northwest. Northwest Nat 72:43–60.

Herrmann J, Andersson KG (1986) Aluminum impact on respiration of lotic mayflies at low pH. Water Air Soil Pollut 30:703–709.

Herrmann R, Klemm K, Tacken E (1989) Behaviour of aluminium species during snowmelt, both downstream and after mixing with nonacidic waters. Aqua Fenn 19:87–94.

Holtze KE, Hutchinson NJ (1989) Lethality of low pH and Al to early life stages of six fish species inhabiting PreCambrian Shield waters in Ontario. Can J Fish Aquat Sci 46:1188–1202.

Horne MT, Dunson WA (1994) Exclusion of the Jefferson salamander, *Ambystoma jeffersonianum,* from some potential breeding ponds in Pennsylvania: effects of pH, temperature, and metals on embryonic development. Arch Environ Contam Toxicol 27:323–330.

Hunn JB, Cleveland L, Little EE (1987) Influence of pH and aluminum on developing brook trout in a low calcium water. Environ Pollut 43:63–73.

Hussein AS, Cantor AH, Johnson TH (1988) Use of high levels of dietary aluminum and zinc for inducing pauses in egg production in Japanese quail. Poult Sci 67: 1157–1165.

Hussein AS, Cantor AH, Johnson TH (1989) Comparison of the use of dietary aluminum with the use of feed restriction for force-molting laying hens. Poult Sci 68:891–896.

Hutchinson NJ, Holtze KE, Munro JR, Pawson TW (1989) Modifying effects of life stage, ionic strength and post-exposure mortality on lethality of $H^+$ and Al to lake trout and brook trout. Aquat Toxicol 15:1–26.

Hutchinson TC, Bozic L, Munoz-Vega G (1986) Responses of five species of conifer seedlings to aluminum stress. Water Air Soil Pollut 31:283–294.

Ingersoll CG, Mount DR, Gulley DD, LaPoint TW, Bergman HL (1990a) Effects of pH, aluminum, and calcium on survival and growth of eggs and fry of brook trout (*Salvelinus fontinalis*). Can J Fish Aquat Sci 47:1580–1592.

Ingersoll CG, Gulley DD, Mount DR, Mueller ME, Fernandez JD, Hockett JR, Bergman HL (1990b) Aluminum and acid toxicity to two strains of brook trout (*Salvelinus fontinalis*). Can J Fish Aquat Sci 47:1641–1648.

Jackson ST, Charles DF (1988) Aquatic macrophytes in Adirondack (New York) lakes: patterns of species composition in relation to environment. Can J Bot 66: 1449–1460.

Jensen FB, Malte H (1990) Acid-base and electrolyte regulation, and haemolymph gas transport in crayfish, *Astacus astacus*, exposed to soft, acid water with and without aluminum. J Comp Physiol B 160:483–490.

Kahl JS, Norton SA, Cronan CS, Fernandez IJ, Bacon LC, Haines TA (1991) Maine. In: Charles DF (ed) Acidic Deposition and Aquatic Ecosystems. Springer-Verlag, New York, pp 203–235.

Kane DA, Rabeni CF (1987) Effects of aluminum and pH on the early life stages of smallmouth bass (*Micropterus dolomieui*). Water Resour Res 21:633–639.

Karlsson-Norrgren L, Dickson W, Ljungberg O, Runn P (1986a) Acid water and aluminium exposure: gill lesions and aluminium accumulation in farmed brown trout, *Salmo trutta* L. J Fish Dis 9:1–9.

Karlsson-Norrgren L, Bjorklund I, Ljungberg O, Runn P (1986b) Acid water and aluminium exposure: experimentally induced gill lesions in brown trout, *Salmo trutta* L. J Fish Dis 9:11–25.

Keller W, Yan ND (1991) Recovery of crustacean zooplankton species richness in Sudbury area lakes following water quality improvements. Can J Fish Aquat Sci 48:1635–1644.

King SO, Mach CE, Brezonik PL (1992) Changes in trace metal concentrations in lake water and biota during experimental acidification of Little Rock Lake, Wisconsin, USA. Environ Pollut 78:9–18.

Klauda RJ, Palmer RE (1987) Responses of blueback herring eggs and larvae to pulses of acid and aluminum. Trans Am Fish Soc 116:561–569.

Klauda RJ, Palmer RE, Lenkevich MJ (1987) Sensitivity of early life stages of blueback herring to moderate acidity and aluminum in soft freshwater. Estuaries 10:44–53.

Krahl-Urban B, Papke HE, Peters K, Schimansky C (1990) Forest Decline. U.S. EPA and German Ministry of Research and Technology. U.S. Environmental Protection Agency, Corvallis, OR.

Krantzberg G, Stokes PM (1989) Metal regulation, tolerance, and body burdens in the larvae of the genus *Chironomus*. Can J Fish Aquat Sci 46:389–398.

Krasovskii GN, Vasukovich LY, Chariev OG (1979) Experimental study of biological effects of lead and aluminum following oral administration. Environ Health Perspect 30:47–51.

Krueger GL, Morris TK, Suskind RR, Widner EM (1985) The health effects of aluminum compounds in mammals. CRC Crit Rev Toxicol 13:1–24.

Lamb DS, Bailey GC (1981) Acute and chronic effects of alum to midge larva (Diptera: Chironomidae). Bull Environ Contam Toxicol 27:59–67.

Lawrence GB, Fuller RD, Driscoll CT (1987) Release of aluminum following whole-tree harvesting at the Hubbard Brook Experimental Forest, New Hampshire. J Environ Qual 16:383–390.

Lee DS, Gilbert CR, Hocutt CH, Jenkins RE, McAllister DE, Stauffer JR Jr (1980) Atlas of North American Freshwater Fishes. North Carolina Biol Surv Publ 1980–12. North Carolina State Museum of Natural History, Raleigh.

Lehtonen J (1989) Effects of acidification on the metal levels in aquatic macrophytes in Espoo, S. Finland. Ann Bot Fenn 26:39–50.

Lundholm CE, Mathson K (1986) Effect of some metal compounds on the $Ca^{2+}$ binding and $Ca^{2+}$-$Mg^{2+}$-ATPase activity of eggshell gland mucosa homogenate from the domestic fowl. Acta Pharmacol Toxicol 59:410–415.

Mackie GL (1989) Tolerances of five benthic invertebrates to hydrogen ions and metals (Cd, Pb, Al). Arch Environ Contam Toxicol 18:215–223.

Madigosky SR, Alvarez-Hernandez X, Glass J (1991) Lead, cadmium, and aluminum accumulation in the red swamp crayfish *Procambarus clarkii* G. collected from roadside drainage ditches in Louisiana. Arch Environ Contam Toxicol 20: 253–258.

Malley DF, Chang PSS (1985) Effects of aluminum and acid on calcium uptake by the crayfish *Orconectes virilis*. Arch Environ Contam Toxicol 14:739–747.

Malte H (1986) Effects of aluminium in hard, acid water on metabolic rate, blood gas tensions and ionic status in the rainbow trout. J Fish Biol 29:187–198.

Marcus MD (1988) Differences in pre- and post-treatment water qualities for twenty limed lakes. Water Air Soil Pollut 41:279–291.

Matsumoto H, Hirasawa E, Takahashi E (1976) Localization of aluminum in tea leaves. Plant Cell Physiol 17:627–631.

Matuszek JE, Beggs GL (1988) Fish species richness in relation to lake area, pH, and other abiotic factors in Ontario lakes. Can J Fish Aquat Sci 45:1931–1941.

McCahon CP, Brown AF, Poulton MJ, Pascoe D (1989) Effects of acid, aluminium and lime additions on fish and invertebrates in a chronically acid Welsh stream. Water Air Soil Pollut 45:345–359.

McCahon CP, Pascoe D (1989) Short-term experimental acidification of a Welsh stream: toxicity of different forms of aluminium at low pH to fish and invertebrates. Arch Environ Contam Toxicol 18:233–242.

McCauley DJ, Brooke LT, Call DJ, Lindberg CA (1986) Acute and Chronic Toxicity of Aluminum to *Ceriodaphnia dubia* at Various pH's. University of Wisconsin, Superior.

McCormick JH, Jensen KM, Anderson LE (1989) Chronic effects of low pH and elevated aluminum on survival, maturation, spawning and embryo-larval development of the fathead minnow in soft water. Water Air Soil Pollut 43:293–307.

McDonald DG, Wood CM, Rhem RG, Mueller ME, Mount DR, Bergman HL (1991) Nature and time course of acclimation to aluminum in juvenile brook trout (*Salvelinus fontinalis*). I. Physiology. Can J Fish Aquat Sci 48:2006–2015.

McKee MJ, Knowles CO, Buckler DR (1989) Effects of aluminum on the biochemical composition of Atlantic salmon. Arch Environ Contam Toxicol 18: 243–248.

Miles AK, Grue CE, Pendelton GW, Soares JH Jr (1993) Effects of dietary aluminum, calcium, and phosphorus on egg and bone of European starlings. Arch Environ Contam Toxicol 24:206–212.

Miller GE, Wile I, Hitchin GG (1983) Patterns of accumulation of selected metals in members of the soft-water macrophyte flora of central Ontario lakes. Aquat Bot 15:53–64.

Mount DR, Ingersoll CG, Gulley DD, Fernandez JD, LaPoint TW, Bergman HL (1988) Effect of long-term exposure to acid, aluminum, and low calcium on adult brook trout (*Salvelinus fontinalis*). 1. Survival, growth, fecundity, and progeny survival. Can J Fish Aquat Sci 45:1623–1632.

Mount DR, Swanson MJ, Breck JE, Farag AM, Bergman HL (1990) Responses of brook trout (*Salvelinus fontinalis*) fry to fluctuating acid, aluminum, and low calcium exposure. Can J Fish Aquat Sci 47:1623–1630.

Mueller CS, Thompson RL, Ramelow GJ, Beck JN, Langley MP, Young JC, Casserly DM (1987) Distributions of Al, V, and Mn in lichens across Calcasieu Parish, Louisiana. Water Air Soil Pollut 33:155–164.

Muniz IP, Leivestad H (1980) Toxic effects of aluminium on the brown trout: *Salmo trutta* L. In: Drablos D, Tollan A (eds) Proceedings of an international conference on the ecological impact of acid precipitation. SNSF Project: Effects on Forest and Fish, Oslo, Norway, pp 320–321.

Muramoto S (1981) Influence of complexans (NTA, EDTA) on the toxicity of aluminum chloride and sulfate to fish at high concentrations. Bull Environ Contam Toxicol 27:221–225.

Nalewajko C, Paul B (1985) Effects of manipulations of aluminum concentrations and pH on phosphate uptake and photosynthesis of planktonic communities in two Precambrian Shield lakes. Can J Fish Aquat Sci 42:1946–1953.

National Academy of Sciences (1980) Mineral tolerance of domestic animals. National Research Council, National Academy of Science, Washington, DC.

National Acid Precipitation Program (NAPAP) (1991) Integrated assessment report. Office of the Director, NAPAP, Washington, DC.

Neville CM (1985) Physiological response of juvenile rainbow trout, *Salmo gairdneri*, to acid and aluminum – prediction of field responses from laboratory data. Can J Fish Aquat Sci 42:2004–2019.

Neville CM, Campbell PGC (1988) Possible mechanisms of aluminum toxicity in a dilute, acidic environment to fingerlings and older life stages of salmonids. Water Air Soil Pollut 42:311–327.

Nosko P, Brassard P, Kramer JR, Kershaw KA (1988) The effect of aluminum on seed germination and early seedling establishment, growth, and respiration of white spruce (*Picea glauca*). Can J Bot 66:2305–2310.

Nuorteva P (1990) Metal distribution patterns and forest decline. Seeking Achilles' heels for metals in Finnish forest biocoenoses, Vol. 11. Publications Department, Environmental Conservation, University of Helsinki, Helsinki, Finland, pp 1–77.

Nyholm NEI, Myhrberg HE (1977) Severe eggshell defects and impaired reproductive capacity in small passerines in Swedish Lapland. Oikos 29:336–341.

Nyholm NEI (1981) Evidence of involvement of aluminum in causation of defective formation of eggshells and of impaired breeding in wild passerine birds. Environ Res 26:363–371.

Ohlendorf HM, Marois KC, Lowe RW, Harvey TE, Kelly PR (1991) Trace elements and organochlorines in surf scoters from San Francisco Bay, 1985. Environ Monit Assess 18:105–122.

Olem H (1991) Liming Acidic Surface Waters. Lewis Publishers, Chelsea, MI.

Ondriecka R, Ginter E, Kortus J (1966) Chronic toxicity of aluminium in rats and mice and its effects on phosphorus metabolism. Br J Indust Med 23:305–312.

Ormerod SJ, Wade KR, Gee AS (1987a) Macro-floral assemblages in upland Welsh streams in relation to acidity, and their importance to invertebrates. Freshwater Biol 18:545–557.

Ormerod SJ, Boole P, McCahon CP, Weatherley NS, Pascoe D, Edwards RW (1987b) Short-term experimental acidification of a Welsh stream: comparing the biological effects of hydrogen ions and aluminium. Freshwater Biol 17:341–356.

Ormerod SJ, Bull KR, Cummins CP, Tyler SJ, Vickery JA (1988) Egg mass and shell thickness in dippers *Cinclus cinclus* in relation to stream acidity in Wales and Scotland. Environ Pollut 55:107–121.

Palmer RE, Klauda RJ, Lewis TE (1988) Comparative sensitivities of bluegill, channel catfish, and fathead minnow to pH and aluminum. Environ Toxicol Chem 7: 505–516.

Parent L, Couture P, Campbell PGC, Dubreuil B (1988) Sensibilite des alevins vesicules du saumon Atlantique a l'acidite en presence et en absence d'aluminium inorganique. Water Pollut Res J Can 23:227–242.

Parker DR, Zelazny LW, Kinraide TB (1989) Chemical speciation and plant toxicity of aqueous aluminum. In: Lewis TE (ed) Environmental Chemistry and Toxicology of Aluminum. Lewis Publishers, Chelsea, MI, pp 117–145.

Parkhurst, BR, Bergman HL, Fernandez J, Gulley DD, Hockett JR, Sanchez DA

(1990) Inorganic monomeric aluminum and pH as predictors of acidic water toxicity to brook trout (*Salvelinus fontinalis*). Can J Fish Aquat Sci 47:1631–1640.

Pauwels SJ, Haines TA (1986) Fish species distribution in relation to water chemistry in selected Maine lakes. Water Air Soil Pollut 30:477–488.

Peterson RH, Bourbonniere RA, Lacroix GL, Martin-Robichaud DJ, Takats P, Brun G (1989) Responses of Atlantic salmon (*Salmo salar*) alevins to dissolved organic carbon and dissolved aluminum at low pH. Water Air Soil Pollut 46: 399–413.

Pillsbury RW, Kingston JC (1990) The pH-independent effect of aluminum on cultures of phytoplankton from an acidic Wisconsin lake. Hydrobiologia 194: 225–233.

Playle RC, Goss GG, Wood CM (1989) Physiological disturbances in rainbow trout (*Salmo gairdneri*) during acid and aluminum exposures in soft water of two calcium concentrations. Can J Zool 67:314–324.

Playle RC, Wood CM (1991) Mechanisms of aluminium extraction and accumulation at the gills of rainbow trout, *Oncorhynchus mykiss* (Walbaum), in acidic soft water. J Fish Biol 38:791–805.

Pollman CD, Canfield DE Jr (1991) Florida. In: Charles DF (ed) Acidic Deposition and Aquatic Ecosystems. Springer-Verlag, New York, pp 367–416.

Ramamoorthy S (1988) Effect of pH on speciation and toxicity of aluminum to rainbow trout (*Salmo gairdneri*). Can J Fish Aquat Sci 45:634–642.

Reader JP, Dalziel TRK, Morris R, Sayer MDJ, Dempsey CH (1991) Episodic exposure to acid and aluminium in soft water: survival and recovery of brown trout, *Salmo trutta* L. J Fish Biol 39:181–196.

Reid SD, McDonald DG, Rhem RR (1991) Acclimation to sublethal aluminum: Modifications of metal-gill surface interactions of juvenile rainbow trout (*Oncorhynchus mykiss*). Can J Fish Aquat Sci 48:1996–2005.

Rengel Z (1992) Role of calcium in aluminium toxicity. New Phytol 121:499–513.

Roberts DA, Singer R, Boylen CW (1985) The submersed macrophyte communities of Adirondack lakes (New York, U.S.A.) of varying degrees of acidity. Aquat Bot 21:219–235.

Robinson DL, Hemkes OJ, Kemp A (1984) Relationships among forage aluminum levels, soil contamination on forages, and availability of elements to dairy cows. Neth J Agric Sci 32:73–80.

Rosa V, Henry PR, Ammerman CB (1982) Interrelationship of dietary phosphorus, aluminum and iron on performance and tissue mineral composition in lambs. J Anim Sci 55:1231–1240.

Rosseland BO, Skogheim OK (1986) Neutralization of acidic brook-water using a shell-sand filter or sea-water: effects on eggs, alevins and smolts of salmonids. Aquaculture 58:99–110.

Rosseland BO, Blaker IA, Bulgar A, Kroglund F, Kvellstad A, Lydersen E, Oughton DH, Salbu B, Staurnes M, Vogt R (1992) The mixing zone between limed and acidic river waters: complex aluminium chemistry and extreme toxicity for salmonids. Environ Pollut 78:3–8.

Sadler K, Lynam S (1985) The mineral content of some freshwater invertebrates in relation to stream pH and calcium concentration. Central Electric Research Laboratory, Leatherland, Surrey, UK.

Sadler K, Turnpenny AWH (1986) Field and laboratory studies of exposures of brown trout to acid waters. Water Air Soil Pollut 30:593–599.

Sadler K, Lynam S (1988) The influence of calcium on aluminium-induced changes in the growth rate and mortality of brown trout, *Salmo trutta* L. J Fish Biol 33: 171–179.

Santlemann MV, Gorham E (1988) The influence of airborne road dust on the chemistry of *Sphagnum* mosses. J Ecol 76:1219–1231.

Sayer MDJ, Reader JP, Morris R (1991a) Embryonic and larval development of brown trout, *Salmo trutta* L.: exposure to aluminium, copper, lead, or zinc in soft, acid water. J Fish Biol 38:431–455.

Sayer MDJ, Reader JP, Morris R (1991b) Embryonic and larval development of brown trout, *Salmo trutta* L.: exposure to trace metal mixtures in soft water. J Fish Biol 38:773–787.

Scanlon P (1990) Effects of acid precipitation on wild mammals and waterfowl. In: Baker JP (ed) Acid deposition: state of the science and technology report 13. Biological effects of changes in surface water acid-base chemistry. NAPAP Publ 3609. Environmental Science Division, Oak Ridge National Laboratory. Office of Director, National Acid Precipitation Program, Washington DC, pp 151–156.

Schaefer DA, Driscoll CT Jr, Dreason RV, Yatsko CP (1990) The episodic acidification of Adirondack lakes during snowmelt. Water Resour Res 26:1639–1647.

Scheuhammer AM (1991) Effects of acidification on the availability of toxic metals and calcium to wild birds and mammals. Environ Pollut 71:329–375.

Schier GA (1985) Response of red spruce and balsam fir seedlings to aluminum toxicity in nutrient solutions. Can J For Res 15:29–33.

Schofield CL, Trojnar JR (1980) Aluminum toxicity to brook trout (*Salvelinus fontinalis*) in acidified waters. In: Toribara T, Miller M, Marrow P (eds) Polluted Rain. Plenum Press, New York, pp 341–363.

Segner H, Marthaler R, Linnenbach M (1988) Growth, aluminium uptake and mucous cell morphometrics of early life stages of brown trout, *Salmo trutta*, in low pH water. Environ Biol Fish 21:153–159.

Short TM, Black JA, Birge WJ (1990) Effects of acid-mine drainage on the chemical and biological character of an alkaline headwater stream. Arch Environ Contam Toxicol 19:241–248.

Siddens LK, Seim WK, Curtis LR, Chapman GA (1986) Comparison of continuous and episodic exposure to acidic, aluminum-contaminated waters of brook trout (*Salvelinus fontinalis*). Can J Fish Aquat Sci 43:2036–2040.

Skogheim OK, Rosseland BO (1984) A comparative study on salmonid fish species in acid aluminium-rich water. I. Mortality of eggs and alevins. National Swedish Board of Fisheries, Institute of Freshwater Research, Drottningholm, Sweden Rept. 61, pp 177–185.

Skogheim OK, Rosseland BO, Sevaldrud IH (1984) Deaths of spawners of Atlantic salmon (*Salmo salar* L.) in River Ogna, SW Norway, caused by acidified aluminium-rich water, Vol. 61. National Swedish Board of Fisheries, Institute of Freshwater Research, Drottningholm, Sweden, pp 195–202.

Skogheim OK, Rosseland BO (1986) Mortality of smolt of Atlantic salmon, *Salmo salar* L., at low levels of aluminium in acidic softwater. Bull Environ Contam Toxicol 37:258–265.

Skogheim OK, Rosseland BO, Hoell E, Kroglund F (1986) Base additions to flowing acidic water: effects on smolts of Atlantic salmon (*Salmo salar* L.). Water Air Soil Pollut 30:587–592.

Skogheim OK, Rosseland BO, Kroglund F, Hagenlund G (1987) Addition of NaOH, limestone slurry and finegrained limestone to acidified lake water and the effects on smolts of Atlantic salmon (*Salmo salar* L.). Water Res 21:435–443.

Sparling DW (1990) Acid precipitation and food quality: inhibition of growth and survival in black ducks and mallards by dietary aluminum, calcium, and phosphorus. Arch Environ Contam Toxicol 19:457–463.

Sparling DW (1991) Acid precipitation and food quality: effects of dietary Al, Ca, and P on bone and liver characteristics in American black ducks and mallards. Arch Environ Contam Toxicol 21:281–288.

Sparling DW, Lowe TP (1996) Metal concentrations of tadpoles in experimental aquatic macrocosms. Environ Pollut (in press).

Spidso TK, Staurnes M (1991) Effects of aluminium on capercaillie *Tetrao urogallus* chicks. Trans Congr Int Union Game Biol 20:500–505.

Sprenger M, McIntosh A, Lewis T (1987) Variability in concentrations of selected trace elements in water and sediment of six acidic lakes. Arch Environ Contam Toxicol 16:383–390.

Sprenger M, McIntosh A (1989) Relationship between concentrations of aluminum, cadmium, lead, and zinc in water, sediments, and aquatic macrophytes in six acidic lakes. Arch Environ Contam Toxicol 18:225–231.

Steiner KC, Barbour JR, McCormick LH (1984) Response of *Populus* hybrids to aluminum toxicity. For Sci 30:404–410.

Stoner JH, Gee AS, Wade KR (1984) The effects of acidification on the ecology of streams in the Upper Tywi Catchment in West Wales. Environ Pollut 35:125–157.

Tan B, Coler RA (1986) Effects of coal pile leachate on Taylor Brook in western Massachusetts. Environ Toxicol Chem 5:897–903.

Taylor GJ (1988) The physiology of aluminum phytotoxicity. In: Sigel H, Sigel A (eds) Metal Ions in Biological Systems, Vol. 24. Aluminum and Its Role in Biology. Marcel Dekker, New York, pp 123–164.

Taylor RW, Ibeabuchi IO, Sistani KR, Shuford JW (1992) Accumulation of some metals by legumes and their extractability from acid mine spoils. J Environ Qual 21:176–180.

Thornton FC, Schaedle M, Raynal DJ (1986) Effect of aluminum on the growth of sugar maple in solution culture. Can J For Res 6:892–896.

Valdivia-Rodriguez RJ (1977) Effect of dietary aluminum on phosphorus utilization by ruminants. Ph.D. dissertation, University of Florida, Gainesville.

Valdivia R, Ammerman CB, Wilcox CJ, Henry PR (1978) Effect of dietary aluminum on animal performance and tissue mineral levels in growing steers. J Anim Sci 47:1351–1356.

Valdivia R, Ammerman CB, Henry PR, Feaster JP, Wilcox CJ (1982) Effect of dietary aluminum and phosphorus on performance, phosphorus utilization and tissue mineral composition in sheep. J Anim Sci 55:402–410.

Vilkka L, Aula I, Nuorteva P (1990) Comparison of the levels of some metals in roots and needles of *Pinus sylvestris* in urban and rural environments at two times in the growing season. Ann Bot Fenn 27:53–57.

Walker RL, Wood CM, Bergman HL (1991) Effects of long-term preexposure to sublethal concentrations of acid and aluminum on the ventilatory response to aluminum challenge in brook trout (*Salvelinus fontinalis*). Can J Fish Aquat Sci 48:1989–1995.

Weatherley NS, Ormerod SJ, Thomas SP, Edwards RW (1988) The response of macroinvertebrates to experimental episodes of low pH with different forms of aluminum, during a natural spate. Hydrobiologia 169:225–232.

Weatherley NS, Rutt GP, Thomas SP, Ormerod SJ (1991) Liming acid streams: aluminium toxicity to fish in mixing zone. Water Air Soil Pollut 55:345–353.

Wile I, Miller G (1983) The macrophyte flora of 46 acidified and acid-sensitive soft water lakes in Ontario. Report, Water Resources Branch, Ontario Ministry of the Environment, Toronto, Canada.

Wisser LA, Heinrichs BS, Leach RM (1990) Effect of aluminum on performance and mineral metabolism in young chicks and laying hens. J Nutr 120:493–498.

Witters HE (1986) Acute acid exposure of rainbow trout, *Salmo gairdneri* Richardson: effects of aluminium and calcium on ion balance and haematology. Aquat Toxicol 8:197–210.

Witters HE, Vangenechten JHD, Van Puymbroeck S, Vandergorght OLJ (1987) Ionoregulatory and haematological responses of rainbow trout *Salmo gairdneri* Richardson to chronic acid and aluminium stress. Ann Soc R Zool Belg 117:411–420.

Witters HE, Van Puymbroeck S, Van Den Sande I, Vanderborght OLJ (1990a) Haematological disturbances and osmotic shifts in rainbow trout, *Oncorhynchus mykiss* (Walbaum) under acid and aluminium exposure. J Comp Physiol B: 160: 563–571.

Witters HE, Van Puymbroeck S, Vangenechten JHD, Vanderborght OLJ (1990b) The effect of humic substances on the toxicity of aluminium to adult rainbow trout, *Onchorhynchus mykiss* (Walbaum). J Fish Biol 37:43–53.

Witters HE, Van Puymbroeck S, Vanderborght OLJ (1991) Adrenergic response to physiological disturbances in rainbow trout, *Oncorhynchus mykiss*, exposed to aluminum at acid pH. Can J Fish Aquat Sci 48:414–420.

Wolff BG, Phillips RE (1990) Effects of dietary aluminum on reproduction in Japanese quail, *Coturnix coturnix japonica*. Water Air Soil Pollut 50:293–299.

Wood CM, Playle RC, Simons BP, Goss GG, McDonald DG (1988a) Blood gases, acid-base status, ions, and hematology in adult brook trout (*Salvelinus fontinalis*) under acid/aluminum exposure. Can J Fish Aquat Sci 45:1575–1586.

Wood CM, McDonald DG, Booth CE, Simons BP, Ingersoll CG, Bergman HL (1988b) Physiological evidence of acclimation to acid/aluminum stress in adult brook trout (*Salvelinus fontinalis*). 1. Blood composition and net sodium fluxes. Can J Fish Aquat Sci 45:1587–1596.

Wood CM, Simons BP, Mount DR, Bergman HL (1988c) Physiological evidence of acclimation to acid/aluminum stress in adult brook trout (*Salvelinus fontinalis*). 2. Blood parameters by cannulation. Can J Fish Aquat Sci 45:1597–1605.

Wood CM, McDonald DG, Ingersoll CG, Mount DR, Johannsson OE, Landsberger S, Bergman HL (1990a) Effects of water acidity, calcium, and aluminum on whole body ions of brook trout (*Salvelinus fontinalis*) continuously exposed from fertilization to swim-up: a study by instrumental neutron activation analysis. Can J Fish Aquat Sci 47:1593–1603.

Wood CM, McDonald DG, Ingersoll CG, Mount DR, Johannsson OE, Landsberger S, Bergman HL (1990b) Whole body ions of brook trout (*Salvelinus fontinalis*) alevins: responses of yolk-sac and swim-up stages to water acidity, calcium, and aluminum, and recovery effects. Can J Fish Aquat Sci 47:1604–1615.

Woodward DF, Farag AM, Mueller ME, Little EE, Vertucci FA (1989) Sensitivity of endemic Snake River cutthroat trout to acidity and elevated aluminum. Trans Am Fish Soc 118:630–643.

Wren CD, Stephenson GL (1991) The effect of acidification on the accumulation and toxicity of metals to freshwater invertebrates. Environ Pollut 71:205–241.

Wright RF (1985) Chemistry of Lake Hovvatn, Norway, following liming and re-acidification. Can J Fish Aquat Sci 42:1103–1113.

Wyman RL, Jancola J (1992) Degree and scale of terrestrial acidification and amphibian community structure. J Herpetol 26:392–401.

Yan ND, Dillon PJ (1984) Experimental neutralization of lakes near Sudbury Ontario. In: Nriagu PO (ed) Environmental Impacts of Smelters. John Wiley & Sons, Toronto, Canada, pp. 417–456.

Yokel RA (1982) Hair as an indicator of excessive aluminum exposure. Clin Chem 28:662–665.

Yokel RA (1984) Toxicity of aluminum exposure during lactation to the maternal and suckling rabbit. Toxicol Appl Pharmacol 75:35–43.

Yokel RA (1987) Toxicity of aluminum exposure to the neonatal and immature rabbit. Fundam Appl Toxicol 9:795–806.

Manuscript received March 23, 1995; accepted April 12, 1995.

# Atrazine Retention and Transport in Soils

Liwang Ma* and H.M. Selim*,†

## Contents



## I. Introduction

As the world's need for food, feed, and fiber increases, pesticides become as indispensable in agriculture production as fertilizer. In the United States, 62% of the agricultural land is treated with pesticides, of which 69% is herbicides, 19% insecticides, and 12% fungicides (Pimental et al. 1991). Since 1945, the use of synthetic pesticides in the U.S. has grown 33-fold (Pimental et al. 1991). Of the pesticides used in agriculture, only 0.1% actually reaches target pests; the rest (over 99%) is distributed into the

*Agronomy Department, Sturgis Hall, Louisiana State University, Baton Rouge, LA 70803 U.S.A.
†Corresponding author.


*Reviews of Environmental Contamination and Toxicology, Vol. 145.*

ecosystem (Pimental and Levitan 1986). The latter may cause serious environmental problems, such as groundwater contamination, food contamination, and air pollution (Pimental et al. 1991).

Atrazine (2-chloro-4-ethylamino-6-isopropylamino-*s*-triazine) is a triazine herbicide (Fig. 1) used worldwide since 1952 to control annual weeds in corn and sugarcane such as henbit (*Lamium amplexicaule* L.), annual bluegrass (*Poa annua* L.), and annual sowthistle (*Sonchus oleraceus* L.) (Anonymous 1990; Ashton and Monaco 1991; Frank and Sirons 1985). The U.S. Environmental Protection Agency (USEPA) estimated that between 32,000 and 34,000 metric tons of atrazine were used in U.S. agricultural crop production in 1993, making it the most heavily used of all pesticides (Aspelin 1994).

Subsequent to this extensive use, numerous reports on groundwater and river contaminations have been documented (Helling and Gish 1986; Isensee et al. 1988; Periera and Rostad 1990; Pionke et al. 1988). A survey of wells in Wisconsin showed contamination of 4 of 13 wells under normal agricultural practice (WNDR 1988). Burkart and Kolpin (1993) found that 1010 of 8699 wells across the U.S. were contaminated with atrazine. During a 5-yr monitoring of tile drains installed at 1.2 m depth in a clay loam soil,

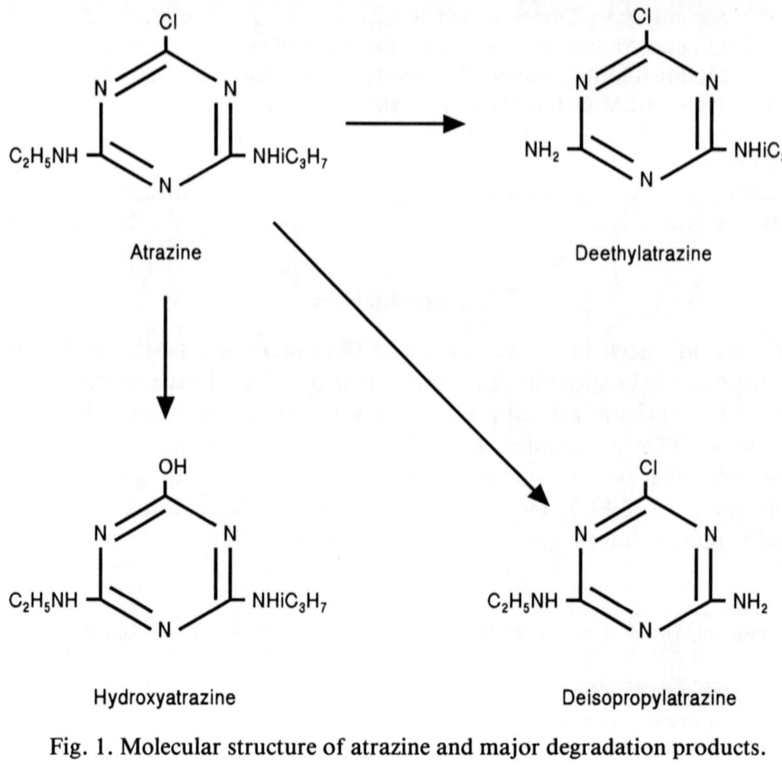


Fig. 1. Molecular structure of atrazine and major degradation products.

Buhler et al. (1993) detected atrazine in 97% of water samples. Southwick et al. (1992) detected atrazine concentrations as high as 403 $\mu$g/L in 1.0-m-depth tile-drain water samples 7 d post-application in a Sharkey clay soil. Smith et al. (1990) observed atrazine at 350 $\mu$g/L 12 d after application in a sandy soil. These reported concentrations are much higher than the safe drinking water lifetime advisory level of 3 $\mu$g/L recommended by the USEPA (USEPA 1991). Although its concentration alone may not exceed the advisory level in some cases, total atrazine and its degradation products can be higher than 3 $\mu$g/L in groundwater samples (Burkart and Kolpin 1993; Jayachandran et al. 1994), rivers (Schottler et al. 1994), precipitation (Nations and Hallberg 1992), and streams and lakes (Spalding et al. 1994).

These widely reported contaminations led to increased public pressure to regulate atrazine application due to its potential adverse effects on human health. Iowa, the first state to regulate atrazine, has limited its use on certain soils. This practice would reduce the allowable application of atrazine by at least 25% statewide (Anonymous 1990). Therefore, studies of atrazine and its role in soils and the environment have been the focus of investigators in several scientific disciplines over the past 40 yr. As early as 1970, a special issue of *Residue Reviews* (Vol. 32) on *s*-triazines was compiled. Moreover, numerous studies have been presented in the literature and scientific meetings related to the environment and agriculture (Cheng 1990; Garner et al. 1986; Somasundaram and Coats 1991). This review provides an overview on atrazine studies from adsorption–desorption and degradation to transport and modeling.

## II. Adsorption and Desorption

### A. Adsorption Isotherms

Atrazine adsorption at equilibrium is usually expressed as adsorption isotherms. The two commonly used isotherms are the Freundlich (Albanis et al. 1989; Brouwer et al. 1990; Clay and Koskinen 1990a,b) and the Langmuir (Ma and Selim 1994a,b; Wang et al. 1992; Weber 1993). The Freundlich isotherm equation may be expressed as

$$S = K_f C^N, \tag{1}$$

and the Langmuir isotherm equation can be written as

$$S = \frac{S_T C}{K_L + C}, \tag{2}$$

where $S$ is the amount sorbed at equilibrium ($\mu$g/g), $C$ is the equilibrium concentration of atrazine in the liquid phase ($\mu$g/mL), $K_f$ is the Freundlich coefficient (mL/g), and $N$ is the Freundlich exponent (dimensionless). The term $S_T$ is atrazine adsorption capacity ($\mu$g/g), and $K_L$ is an atrazine affinity parameter ($\mu$g/mL).

The Freundlich equation is perhaps the most commonly used isotherm in pesticide studies, although it is inferior to the Langmuir equation in some studies (Ma and Selim 1994a,b; Wang et al. 1992). The value of $N$ for atrazine is usually less than or close to 1.0 (Clay and Koskinen 1990a,b; Ma et al. 1993). When $N = 1$, Equation (1) is reduced to the linear isotherm equation:

$$S = K_d C, \tag{3}$$

where $K_d$ is the distribution coefficient (mL/g), which is dependent not only on pesticide characteristics but also on soil properties.

Two methods have been used to obtain pesticide adsorption isotherms in the laboratory (Green and Corey 1971; Green et al. 1980; Hance 1988; Johnson and Farmer 1993); the batch method and the flow method. The batch method was carried out by suspending a certain amount of soil in a fixed volume of pesticide solution. The flow method, on the other hand, involves continuous flow of a pesticide solution of known concentration into a soil column. The soil column (with or without stirring) can be as thin as a few millimeters or as thick as 20 cm or more. The amount of pesticide adsorbed is usually calculated by its disappearance from the liquid phase. Pesticide adsorption can also be estimated from the amount of pesticide displaced from the soil column with pesticide-free solvent (Green and Corey 1971). Other methods of obtaining equilibrium adsorption constants using the flow method are the retardation method (Elrick et al. 1966; Johnson and Farmer 1993) and the breakthrough maxima method (Kookana et al. 1992a,b).

Pesticide adsorption reaches equilibrium sooner in batch experiments than in flow systems, which was attributed to the mechanical shaking, vortexing, and centrifuging involved in batch experiments (Kookana et al. 1992b). The sorption coefficients obtained from the continuous flow method are significantly lower than those calculated from batch experiments (Davidson et al. 1968; Kookana et al. 1992a). Therefore, the validity of the adsorption–desorption isotherms obtained using the batch technique is questionable in miscible column experiments. Rao and Jessup (1983) suggested that the batch method was insensitive and might be unsuitable for investigating sorption kinetics.

Indirect measurement of atrazine adsorption by its disappearance from liquid solution may exclude degradation and volatilization loss during experiments (Clay et al. 1988; Singh et al. 1990a). Singh et al. (1990a) observed that $K_f$ calculated from atrazine disappearance in the soil solution is as much as 2.5 times higher than that based on directly measured atrazine on the soil phase. However, if loss from degradation and volatilization is small, it is valid to measure atrazine adsorption by atrazine disappearance from the soil solution. The concentration difference method may also not be accurate if the concentration change is relatively small compared to the error in pesticide measurements (Boesten 1990; Green and Yamane 1970; Johnson and Farmer 1993).

## B. Effects of Soil Properties

Atrazine can be adsorbed by means of physical fixation, ionic bond, covalent bond, hydrogen bonding, van der Waals forces, and hydrophobic bonding (Barriuso et al. 1994; Wang et al. 1992; Weber 1993). The importance of each adsorption mechanism depends on soil composition and pH. Atrazine is adsorbed in protonated form by cation exchange at low soil pH and in molecular form through physical adsorption forces at high pH, due to its basic nature (Weber 1993; Wang et al. 1992). Welhouse and Bleam (1993), using nuclear magnetic resonance (NMR), demonstrated the formation of hydrogen-bond complexes between atrazine and compounds commonly found in soil organic matter. They further measured the formation constants of the complexes and found that weak to moderately strong complexes are formed with amine, hydroxyl, and carbonyl functional groups. However, strong complexation is observed with carboxylic acid and amide function groups.

Atrazine adsorption is higher at lower soil pH and reaches its maximum adsorption when surface acidity is approximately equal to the dissociation constant of atrazine ($pK_a = 1.68$) (Kalouskova 1989; Wang et al. 1992; Weber 1993). Soil organic matter has the highest adsorption affinity to atrazine among the soil constituents (Laird et al. 1994; Wang et al. 1992). In a study with a Webster soil, Laird et al. (1994) found that organic matter accounted for only 11% of the clay weight but contributed to 68% of atrazine adsorption. Brouwer et al. (1990) observed a logarithmic increase of atrazine distribution coefficient ($K_d$) with organic matter content. Besides the absolute amount of organic matter, the stage of decomposition is also important in determining atrazine adsorption (Singh et al. 1989). Laird et al. (1994) found that organic matter associated with the coarse clay fraction had higher affinity for atrazine and exhibited greater sorption hysteresis than organic matter associated with the fine clay. They further analyzed the composition of the organic matter and found that organic matter from the coarse fraction contains more carbonyl-containing function groups (carboxyls, ketones, and aldehydes) while that from the fine fraction is enriched with aliphatic C, bound hydroxyls, and amines.

Clay minerals can also make a significant contribution to atrazine adsorption because of its negative surface charge and low pH in the immediate vicinity of the clay surface (Yaron et al. 1985). However, the importance of clay minerals decreases with increase in organic matter content. Because of the different characteristics of clay surface, atrazine adsorption can vary considerably among clay minerals. Gilchrist et al. (1993) found that the adsorption capacity of clays for atrazine followed the order of montmorillonite > illite > kaolinite. Laird et al. (1992) and Barriuso et al. (1994) further showed that montmorillonite from different origins had different affinities for atrazine, with adsorption from 0 to nearly 100%, depending on surface characteristics such as cation exchange capacity and surface charge density (Fig. 2).

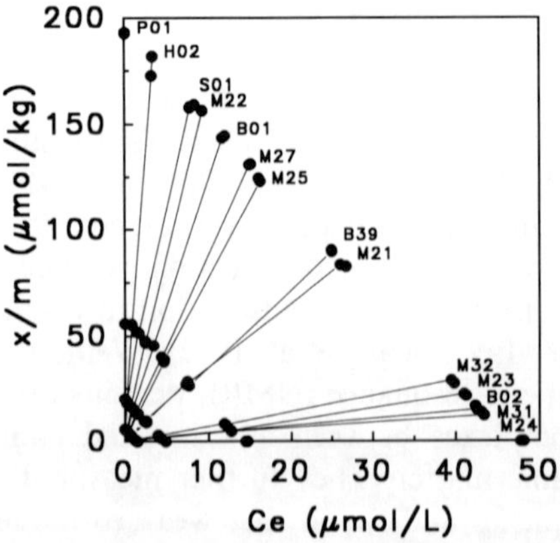


Fig. 2. Freundlich isotherms for adsorption of atrazine on reference and soil smectitic clay samples; atrazine adsorption was determined in duplicate at four levels for each of the Ca-saturated clay samples in 0.01 M $CaCl_2$ solutions. H02, hectorite; B01, B02, M21, M22, M23, M24, M25, M27, M32, B39, montmorillonite; M31, smectite/illite; S01, saponite; P01, beidellite (after Laird et al. 1992).

Soil moisture content or degree of water saturation in the soil has been shown to have an effect on *s*-triazine protonation and adsorption. Grover and Hance (1970) reported a threefold increase of the Freundlich $K_f$ with a decrease in the soil-to-solution ratio from 4 : 1 to 1 : 10. This increase in adsorption was explained by greater dispersion of soil aggregates at a soil : water ratio of 1 : 10. Thus, under field conditions, drying and wetting cycles can significantly alter atrazine adsorption–desorption characteristics (Pignatello et al. 1993). However, in the commonly used soil : water ratio range for laboratory batch experiments, the atrazine adsorption isotherm is not often affected by the soil : water ratio (Dao and Lavy 1978; Ma et al. 1993; Ma and Selim 1994a) (Fig. 3).

The importance of soil constituents to atrazine adsorption in soils may be inferred from studies with isolated materials such as clay (Barriuso et al. 1994; Gilchrist et al. 1993; Laird et al. 1994). and organic matter (Barriuso et al. 1992; Wang et al. 1990, 1992). Wang et al. (1992) were able to predict atrazine adsorption capacity in a Laurentian soil by adding the individual contributions of organic matter and clay minerals. Another way of estimating the importance of soil constituents is to measure the decrease in atrazine adsorption after removing that soil component (Huang et al. 1984; Laird et al. 1994). Laird et al. (1994) separated three clay size fractions (0.2–2.0, 0.02–0.2, and <0.02 μm) and treated each fraction with $H_2O_2$ or $H_2O_2$ + DCB (dithionite citrate bicarbonate). They found that measured distribution coefficients ($K_d$) were significantly different between particle sizes and treatments. However, this technique cannot assess the contribution of a soil component to atrazine adsorption in a real soil system, because organic

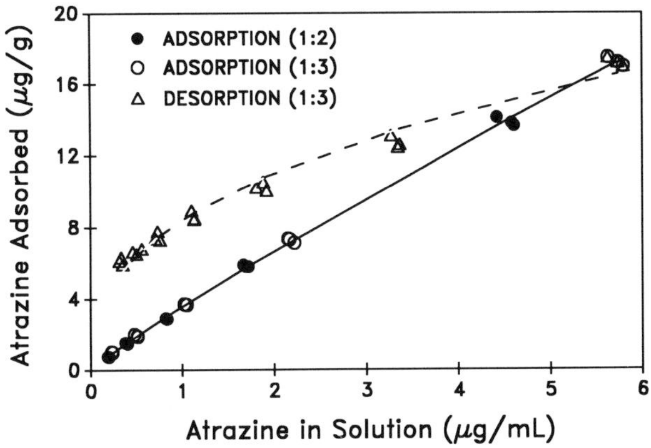


Fig. 3. The effect of soil : water ratio on atrazine adsorption–desorption isotherms for a Sharkey soil. Desorption was conducted after 4 d of incubation between completion of adsorption (*circles*) and initiation of desorption (*triangles*). No significant differences were obtained between soil : water ratios of 1 : 2 and 1 : 3 during adsorption ($p$ = .294). Solid and dashed curves are fitted with the Freundlich equation (after Ma et al. 1993).

matter, sesquioxides, and clay minerals are closely interrelated in soils. These treatments do more than eliminate a constituent, and they affect the remaining materials by blocking or unblocking adsorption sites for atrazine (Calvet 1980; Wang et al. 1992).

Since organic matter is the major determining factor on atrazine adsorption (Brouwer et al. 1990), $K_{oc}$ ($=K_d$/OC, where OC is the percentage of organic carbon content) is perhaps more meaningful in comparing atrazine adsorption among soils. $K_{oc}$ values for atrazine in surface soils have been reported as 160 mL/g (Jury et al. 1987), 107 mL/g (Gustafson 1989), 100 mL/g (Wauchope et al. 1992), and 347 mL/g (Johnson et al. 1995). However, $K_{oc}$ may not be a reliable indicator for soils having low organic carbon contents ($\leq 1\%$), in which the contribution from minerals is significant (Roy and Krapac 1994). The high $K_{oc}$ value of Johnson et al. (1995) was due to a significant contribution from the clay fraction of the Sharkey soil. Roy and Krapac (1994) obtained $K_{oc}$ for low organic adsorbents ranging from 112 to 1680 mL/g. They also found that pH was a better indicator of atrazine adsorption in soils with low organic matter content than organic carbon. In a study with two Wisconsin soils, Seybold et al. (1994) obtained $K_{oc}$ values of 62–141 mL/g for prairie soil and 166–369 mL/g for forest soil. Such variability in $K_{oc}$ among soils was attributed to their differences in organic matter composition as well as contributions from clay minerals (Seybold et al. 1994).

Efforts have also been made to estimate $K_{oc}$ from physical and/or chemical properties of pesticides, such as solubility, octanol–water partition coefficient ($K_{ow}$), melting point, and pure solute molar volume (Briggs 1981; Green and Karickhoff 1990; Mingelgrin and Gerstl 1983; Singh et al. 1989). At best, estimated $K_{oc}$ values could be used as first approximations in modeling chemical transport and in establishing a reference database in the literature (Green and Karichhoff 1990; Johnson and Farmer 1993). However, such estimates may be of little practical value in predicting atrazine transport in soils, since atrazine mobility is strongly affected by soil properties (Mingelgrin and Gerstl 1983; Singh et al. 1989).

### C. Desorption Hysteresis

Atrazine can be released from soil adsorption sites when its concentration in the solution phase decreases due to dilution, volatilization, or translocation. This process is called desorption and can be initiated by diluting or replacing the soil solution with atrazine-free solutions (Clay and Koskinen 1990a; Ma et al. 1993; Swanson and Dutt 1973). The desorption isotherm is also described by the Freundlich equation. However, the fitted $N$ is consistently lower and $K_f$ is higher than that obtained from adsorption isotherms (Clay and Koskinen 1990a,b; Laird et al. 1994; Ma et al. 1993; Stehouwer et al. 1993). Therefore, less herbicide is desorbed from soils than predicted by the adsorption isotherm, which is often referred to as hysteresis (Ma et al. 1993; Selim et al. 1976; Swanson and Dutt 1973) (Fig. 3). Laird et al. (1994) compared the hysteresis phenomenon in different clays based on the $K_d$ values obtained from adsorption and desorption isotherms. They found that extensive hysteresis resulted from atrazine adsorption on organic components, while little or no hysteresis was observed for clay minerals.

Swanson and Dutt (1973) found that $N$ for desorption was independent of atrazine initial concentration and that an $N_a/N_d$ ($N_a$ = adsorption; $N_d$ = desorption) value of 2.3 was adequate to describe atrazine adsorption–desorption hysteresis. Barriuso et al. (1994) obtained $N_a/N_d$ values ranging from 0.64 to 2.56 depending on the type of smectites. Ma et al. (1993) quantified atrazine adsorption–desorption hysteresis using the maximum difference between adsorption and desorption isotherms such that

$$\omega = \frac{\mathrm{Max}(S_d - S_a)}{S_a} \times 100, \tag{4}$$

where $\omega$ is a hysteresis index (dimensionless), and $S_a$ and $S_d$ are the amounts of atrazine retained during the adsorption and desorption processes ($\mu$g/g), respectively. When the Freundlich equation is applied, Equation (4) can be simplified to

$$\omega = \left(\frac{N_a}{N_d} - 1\right) \times 100. \tag{5}$$

As shown in Fig. 4, Ma et al. (1993) found that $\omega$ increased with incubation time ($t$), with $N_a/N_d$ ranging from 1.56 at 0 d incubation to 4.52 at 24 d

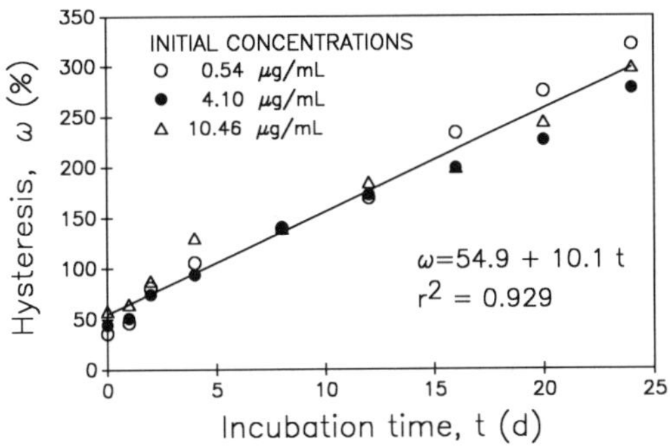


Fig. 4. Calculated hysteresis $\omega$ (percent) versus incubation time (time between the completion of adsorption and initiation of desorption) in a Sharkey soil for several initial concentrations. No significant differences among initial concentrations ($p$ = .690) were obtained (after Ma et al. 1993).

incubation (Fig. 5). The Freundlich $K_f$ and percentage recovery, on the other hand, were dependent on both incubation time and initial (input) atrazine concentration (Figs. 6 and 7).

Hysteresis is more obvious under low soil pH, long reaction time, high organic matter content, frequent drying and wetting, and high degradation rates (Best and Weber 1974; Clay and Koskinen 1990a; Pignatello and Huang 1991). Although there is no universal explanation for the differences between adsorption and desorption isotherms, several causes may be responsible for the observed hysteresis, such as kinetic adsorption (Ma and Selim 1994a; Pignatello and Huang 1991), irreversible conjugation with soil organic components (Stevenson 1976), and degradation (Ma et al. 1993;

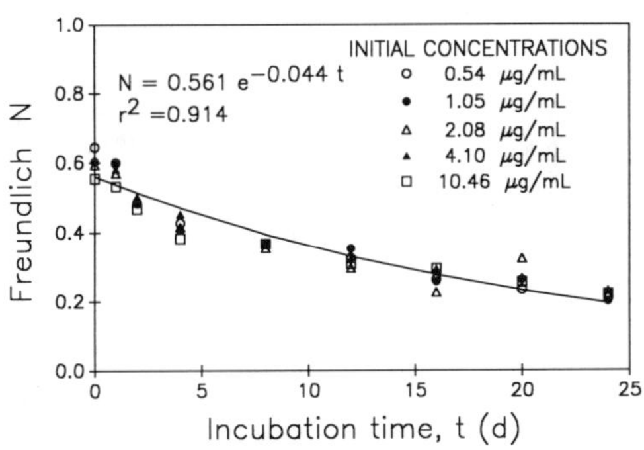


Fig. 5. Fitted Freundlich $N$ for desorption versus incubation time (time between the completion of adsorption and initiation of desorption) in a Sharkey soil for several initial concentrations. No significant differences among initial concentrations ($p$ = .635) were obtained (after Ma et al. 1993).

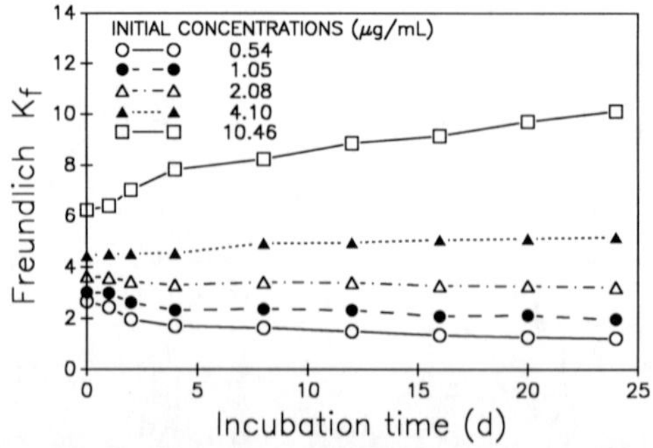


Fig. 6. Fitted Freundlich $K_f$ for desorption versus incubation time (time between the completion of adsorption and initiation of desorption) in a Sharkey soil for several initial concentrations (after Ma et al. 1993).

Singh et al. 1990b). Experimental artifacts may also contribute to the observed hysteresis phenomenon due to changes in soil structure, ionic strength, and dissolved organic carbon content of the solution phase during desorption processes (Barriuso et al. 1992; Bowman and Sans 1985; Gschwend and Wu 1985; Wang et al. 1992).

## D. Bound Residues

Bound residues are defined as the fraction of the total amount of atrazine, initially applied to the soil, that cannot be extracted by methods commonly used in residue analysis, such as organic solvent extractions (Gilchrist et al. 1993). Methanol is the organic solvent most commonly used in atrazine extractions. Atrazine residue can be estimated either by the difference be-

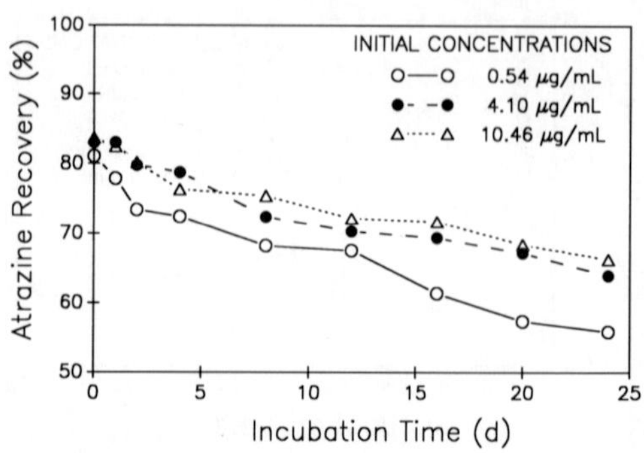


Fig. 7. Total atrazine recovery following six desorptions with 0.01 N $Ca(NO_3)_2$ versus incubation time (the time between the completion of adsorption and initiation of desorption) in a Sharkey soil. Significant differences were obtained among different initial concentrations ($p$ = .039) (after Ma et al. 1993).

tween applied and methanol-recovered atrazine or by oxidizer combustion (Barriuso et al. 1991; Clay et al. 1988; Sorenson et al. 1993; Topp et al. 1994). The amount of residues, in the form of atrazine or its degrades, increases with atrazine–soil reaction time (Sorenson et al. 1993; Topp et al. 1994). Sorenson et al. (1993) found that only 5% of $^{14}C$-atrazine was nonextractable with methanol immediately following application, but up to 18% was nonextractable 16 mon after application in an Estherville sandy loam soil. Barriuso et al. (1991) observed 15%–40% in bound form after 6 mon of incubation in three French soils (clay loam, silt loam, and silty clay), and they further showed that bound residues were closely related to the organic fraction of the soil. Topp et al. (1994) observed a first-order formation of nonextractable residues during a 63-d incubation study with three loam soils. At the end of 63 d, nonextractable residues comprised 45%–60% of the atrazine originally applied to the soils. A large percentage of bound residues was observed with decrease in initial atrazine concentration, which was explained by more adsorption of atrazine to irreversible adsorption sites (Johnson and Farmer 1993; Smith et al. 1992). However, the amount of nonextractable atrazine is not necessarily correlated with the lack of availability to plants, animals, and microorganisms. Small amounts of the nonextractable atrazine remain available to plants and microorganisms (Bertin et al. 1990; Khan 1991).

## III. Degradation

### A. Hydrolysis

Numerous investigations have been carried out on atrazine hydrolysis in soils. Atrazine hydrolysis to hydroxyatrazine (Fig. 1) is considered as a chemical process catalyzed by the soil surface (Albanis et al. 1989; Bacci et al. 1989; Burkhard and Guth 1981; Gamble and Khan 1990, 1992; Wang et al. 1992). In soils, hydrolysis of atrazine is favored by low soil pH, high organic matter content, low moisture content, high temperature, and high clay content (Burkhard and Guth 1981; Gamble and Khan 1988, 1992; Hiltbold and Buchanan 1977; Huang and Pignatello 1990; Li and Felbeck 1972). Hydrolysis is a primary pathway for atrazine degradation (Muir and Baker 1978; Sorenson et al. 1993). In a field study with a sandy loam soil, Sorenson et al. (1993) found that hydroxyatrazine was the predominant degradation product of atrazine. The proportion of $^{14}C$ as hydroxyatrazine in the top 10 cm increased from 15% at 2 mon after application to 37% at 16 mon.

Hydroxyatrazine is adsorbed more strongly in soils than atrazine (Clay et al. 1988; Clay and Koskinen 1990b; Schiavon 1988a,b) because its protonation is higher at the same soil pH (the $pK_a$ of hydroxyatrazine is 5.1 versus 1.68 for atrazine) (Brouwer et al. 1990; Wang et al. 1992). It is necessary to acidify (pH $<1$) the extracting solvent to recover hydroxyatrazine from soils (Baluch et al. 1993; Chung et al. 1994; Steinheimer 1993).

Clay and Koskinen (1990b) studied the adsorption and desorption of atrazine and hydroxyatrazine, and found that, although atrazine showed hysteresis, atrazine can be desorbed to some extent, whereas the desorption of hydroxyatrazine was negligible. Thus, leaching of hydroxyatrazine in soils is not likely (Schiavon 1988a; Somasundaram et al. 1991). Detection of hydroxyatrazine below the soil surface was due to *in situ* degradation of atrazine rather than leaching from the top soil layers (Sorenson et al. 1993).

Hydroxyatrazine does not interfere with atrazine adsorption (Wang et al. 1990, 1992). This independent adsorption behavior indicates that these two compounds are adsorbed on different sites on the soil matrix (Wang et al. 1992). Since the p$K_a$ of atrazine is 1.68 and that of hydroxyatrazine is 5.1, their degrees of protonation are different in soils. Therefore, the competition or interference between the two compounds should be minimal, if not negligible.

### B. Microbial Degradation

Atrazine can be degraded by both fungal and nonfungal microorganisms. Kaufmann and Blake (1970) observed its degradation by 12 different fungi in a basal salt medium supplemented with sucrose. Schocken and Speedie (1984) also observed atrazine degradation by the marine fungus *Periconia prolifica*. Although fungi are the main microorganisms participating in atrazine biodegradation (Kaufmann and Blake 1970; Levanon 1993), atrazine can also be degraded by bacteria belonging to *Nocardia* (Giardina et al. 1980), *Pseudomonas* (Behki and Khan 1986), and *Rhodococcus* (Behki and Khan 1994). Atrazine can be degraded aerobically or anaerobically, and may be used as the sole source of C and N (Mandelbaum et al. 1993; Radosevich et al. 1995). However, the degradation products may vary from microbe to microbe and may also depend on the growth stage of microorganisms (Behki and Khan 1994; Levanon 1993). Fungi are mainly responsible for mineralization of the side chains, while bacteria can actively break the heterocyclic ring (Levanon 1993; Mandelbaum et al. 1993; Radosevich et al. 1995).

Major degradation pathways by microorganisms, including dealkylation and ring cleavage, were reviewed by Erickson and Lee (1989). Deethylatrazine and deisopropylatrazine are the two major microbial dealkylation products in soils (Fig. 1). Both ring cleavage and dealkylation of atrazine and hydroxyatrazine are enhanced by addition of an energy source but are retarded under anaerobic and denitrifying conditions due to high pH, restricted aeration, and fewer microorganisms (Behki and Khan 1994; McMahon et al. 1992; Nair and Schnoor 1992). Microbial degradation of the side chains is more rapid than that of ring cleavage, with the ease of deethylation over deisopropylation (Adams and Thurman 1991; Durand and Barcelo 1992; McMahon et al. 1992; Nair and Schnoor 1992; Wienhold et al. 1993). However, Schocken and Speedie (1984) found that deisopropylatrazine was the predominant biodegradation product over deethylatrazine by a marine fungus, *Periconia prolifica* Anastasiou. Hydroxyatrazine may be an intermediate in atrazine ring mineralization (Mandelbaum et al. 1993).

The microbial degradation products, deethylatrazine and deisopropylatrazine, have higher mobility than hydroxyatrazine (Adams and Thurman 1991; Roy and Krapac 1994; Schiavon 1988b; Sorenson et al. 1993). Compared to atrazine, deethylatrazine and deisopropylatrazine are adsorbed to a lesser degree, and their adsorption is reversible (Brouwer et al. 1990; Roy and Krapac 1994). Sorenson et al. (1993) found that although hydroxyatrazine was the predominant degradation product in the top 10 cm of a sandy loam soil, deethylatrazine was the predominant product at the 10–30-cm depth and accounted for as much as 23% of the $^{14}C$ present in the 10–20-cm soil depth. They also noticed that the proportion of deethylatrazine and deisopropylatrazine increased with depth while that of hydroxyatrazine decreased with soil depth. Gaynor et al. (1995) found significant amounts of deethylatrazine in both surface runoff and tile-drainage samples. Therefore, microbial degradation products have higher potential to contaminate groundwater compared to hydroxyatrazine. Since deethylatrazine is more persistent than deisopropylatrazine (McMahon et al. 1992; Nair and Schnoor 1992), a deethylatrazine-to-atrazine ratio (DAR) was recently proposed as an indicator of atrazine mobility in soils (Adams and Thurman 1991; Jayachandran et al. 1994; Schottler et al. 1994; Spalding et al. 1994). High DAR values indicate a long residue time of atrazine in soils; DAR values less than 0.5 may suggest preferential flow of atrazine in soils.

## C. Dissipation

Dissipation as a first-order kinetic process is often used for both biological and nonbiological degradation (Huang and Pignatello 1990; McMahon et al. 1992; Nair and Schnoor 1992). Therefore, an overall dissipation rate is usually used without distinguishing between biological and nonbiological degradation. Such a pseudo-first-order dissipation rate can be quantified by an apparent rate constant ($\mu$) or a half-life ($t_{1/2} = 0.6932/\mu$). Phogat et al. (1984) found that atrazine degradation proceeded as first-order reactions in both sterilized and unsterilized soils, despite the half-life of 80 and 50 d, respectively. Nash (1988) observed pseudo-first-order kinetics for the overall dissipation of atrazine in a field study (a total loss from application, volatilization, bio- or chemical degradation, adsorption, etc.), with the dissipation rate constants ranging from 0.006 to 0.085 $d^{-1}$. A similar range of dissipation constants was reported by Topp et al. (1994) in a laboratory study with three loam soils. Southwick et al. (1990a, 1992) observed a half-life of 35–36 d in the top 2.5 cm of a Commerce clay loam and 24 d in a Sharkey clay soil. Ghadiri et al. (1984) found that atrazine dissipation was similar in conventional-till and no-till treatments, with half-lives of 42 and 50 d, respectively. Smith and Walker (1989) found that the half-life of atrazine decreased with temperature in a clay soil. They further noticed a decrease of half-life with soil moisture content (up to 26% gravimetric moisture content). A similar soil moisture effect was reported by Singh et al. (1990b). Atrazine half-life was also longer at lower atrazine initial concentrations (Singh et al. 1990b). However, atrazine dissipation in water bodies is slower than that in soils. Elint et al. (1993) found that atrazine was not degraded in groundwater samples during an incubation period of 539 d at 10 °C. Widmer et al. (1993) reported an average half-life of 4.5 yr ± 42 wk in deionized and well water samples stored at 4 °C and 30 °C. They concluded that it might not be necessary to refrigerate water samples prior to analysis.

Gustafson and Holden (1990) showed that first-order kinetics does not always provide adequate description of the overall dissipation rate of atrazine. Richter et al. (1992) found that atrazine degradation may be better described by combining first-order degradation with simultaneous first-order reversible adsorption. Gustafson and Holden (1990) obtained an improved description of pesticide dissipation in soils by considering the overall dissipation process to be a linear combination of several distinct first-order degradation reactions. Their approach can be reduced to a pseudo-first-order model when the soil is spatially uniform and all degradation reactions have the same rate coefficients.

## IV. Measurements

### A. Extraction

Atrazine extraction methods vary from laboratory to laboratory since the procedures are empirical due to lack of knowledge on pesticide–adsorbent, solvent–absorbent, and pesticide–solvent relationships (Chesters et al. 1974). Shaking, reflux, and soxhlet are the methods commonly used in soil extractions. A number of scientists have found that high temperature improved atrazine extraction (Chung et al. 1994; Huang and Pignatello 1990; Steinheimer 1993). The extracting solvents also vary from acid solutions to organic solvents. Atrazine recovery may range from 70% to 90% depending on such factors as soil characteristics, extracting procedures, and solvent. Steinheimer (1993) extracted atrazine and its major degradation products with 0.35 M HCl solution; this method was later used by Chung et al. (1994). Organic solvents used for atrazine extraction include methanol (Best and Weber 1974; Ghadiri et al. 1984), ethylacetate (Fleming et al. 1992; Southwick et al. 1992), dichloromethane (Sauer et al. 1990), and chloroform (White et al. 1967). Mixture of solvents is also used to improve extraction efficiency, such as acidified methanol (Assaf and Turco 1994) and a methanol–water solution (Huang and Pignatello 1990). Such a diversity of extraction methods adds difficulty in comparing data and parameters from different laboratories. In addition, the extraction efficiency of all the methods decreases with atrazine–soil reaction time due to the formation of nonextractable residues (Sorenson et al. 1993; Topp et al. 1994). Soil extracts may need further "cleanup," which can be carried out with liquid-liquid partition (Baluch et al. 1993) or column separation (Chesters et al. 1974; Sauer et al. 1990).

Atrazine extraction from water samples can be done by either liquid-liquid partition or solid-phase extraction (Basta and Olness 1992; Buhler et al. 1993; Fleming et al. 1992). The organic solvents used in the liquid-liquid partition method include dichloromethane (Buhler et al. 1993; Pignatello et al. 1993; Sauer and Daniel 1987) and Haxane (Southwick et al. 1990b). Recently, C-18 solid-phase extraction disks or tubes have been increasingly used for atrazine extraction (Fleming et al. 1992; Gaynor et al. 1995; Shipitalo et al. 1990; Steinheimer 1993). Basta and Olness (1992) recommended inexpensive nonpolar resin as an extracting medium with minimal loss of efficiency. A surrogate pesticide, including terbutylazine (Jayachandran et al. 1994; Spalding et al. 1994; Steinheimer 1993), trifluralin (Gish et al. 1994), and metribuzin (Buhler et al. 1993), is usually added into each sample during solid-phase extractions to check individual sample recovery.

### B. Concentration Determination

The commonly used methods for quantitative measurement are bioassay (Hiltbold and Buchanan 1977; Libik and Romanowski 1976), enzyme-linked immunosorbent assay (ELISA) (Green et al. 1995; Thurman et al. 1990), ultraviolet spectrometry (Kalouskova 1989; Li and Felbeck 1972), thin layer chromatography (TLC) (Alhajjar et al. 1990; Baluch et al. 1993; Topp et al. 1994), high-pressure liquid chromatography (HPLC) (Gamble and Khan 1988; Ma et al. 1993; Steinheimer 1993), and gas chromatography (GC) (Durand and Barcelo 1992; Fleming et al. 1992; Widmer et al. 1993). The choice of methods depends on the availability of instrument, the purpose of the experiment, and the level of accuracy required. Of these methods, TLC, HPLC, and GC are most commonly used.

TLC is commonly used to separate atrazine from its degradation products. Atrazine was identified from its relative mobility or retardation factor ($R_f$), which is the rate of movement of atrazine relative to the rate of movement of the developing solvent. Baluch et al. (1993) differentiated atrazine from its major degradation products using a solvent system of chloroform–methanol–formic acid–water. The developing solvents vary from laboratory to laboratory, including benzene–acetic acid–water (Alhajjar et al. 1990; Clay et al. 1988), butanol–acetic acid–water (Clay and Koskinen 1990a,b), chloroform–methanol (Topp et al. 1994), and chloroform–nitromethane and petroleum ether–chloroform–methanol (Best and Weber 1974).

GC has been widely used in pesticide analysis and is capable of atrazine measurement at the parts per billion (ppb) level. A capillary column is often used in atrazine analysis. Oven temperature is usually programmed to increase resolution (Basta and Olness 1992; Durand and Barcelo 1992; Widmer et al. 1993) but may not be necessary for relatively clean samples (Southwick et al. 1990a). Atrazine can be detected with a nitrogen-phosphorus detector (NPD) (Basta and Olness 1992; Durand and Barcelo 1992; Fleming et al. 1992; Widmer et al. 1993), $^{63}$Ni detector (Sauer et al. 1990), electrolytic conductivity detector (Southwick et al. 1990a), thermionic-specific detector (Shipitalo et al. 1990), and gas chromatography–mass spectrometry (GC–MS) (Pignatello et al. 1993; Thurman et al. 1990).

HPLC has been widely used in recent years in atrazine measurement and is less sensitive than GC. Atrazine is usually separated by a silica-based column and detected with an UV detector. The mobile phase can be methanol–acetonitrile–water (Bouchard 1987), acetonitrile–water (Basta and Olness 1992), or methanol–water (Clay et al. 1988; Ma et al. 1993), and its pH may be adjusted by HCl or ammonium acetate as needed (Assaf and Turco 1994; Gamble and Khan 1988). Column flow rate is usually about 1.0 mL/min and atrazine is detected at a wavelength between 220 nm and 230 nm.

More studies are now conducted with $^{14}$C-labeled atrazine (Ma and Selim 1994a,b; Sorenson et al. 1993). Atrazine concentrations are then analyzed using liquid scintillation counting (LSC). Atrazine can be labeled on the ring and/or on the side chains. It is convenient to study bound residues and degradation pathways using $^{14}$C-atrazine labeled on different positions (Barriuso et al. 1991; Mandelbaum et al. 1993; Radosevich et al. 1995; Topp et al. 1994). The other advantage of using radioisotopes is the direct measurement of adsorbed atrazine in the soil without extraction (Hubbs and Lavy 1990). $^{14}$C counting using LSC cannot differentiate atrazine from some of its degradation products, however.

## V. Transport

Atrazine is applied to a field at 1.7–2.8 kg a.i. ha$^{-1}$ depending on the soil texture (Gaynor et al. 1995; Sauer and Daniel 1987; Steinheimer 1993). The applied atrazine is subject to runoff loss, infiltration loss, volatilization, soil adsorption, and degradation. Because atrazine has a low partition coefficient between the solution and vapor phases (Henry's law constant) (Jury et al. 1983), volatilization loss is negligible (Alhajjar et al. 1990; Bacci et al. 1989). Elling et al. (1987) detected trace amounts of atrazine on a thin-layer chromatograph hanging in the atmosphere for 3 wk after atrazine application. Glotfelty et al. (1989) measured atrazine volatilization losses from a fallow field and found that only 2.4% of the applied atrazine was lost after 24 d. Although Wienhold et al. (1993) reported a 14% loss from volatilization after 35 d under optimum evaporation conditions in a laboratory chamber, atrazine loss under field conditions was in the range of 1%–9% (Glotfelty et al. 1989; Whang et al. 1993; Wienhold and Gish 1994; Wienhold et al. 1993).

Atrazine runoff loss is very small (0.5%–3% of applied) and is not significantly different between tillage practices. The environmental conditions, such as rainfall pattern, are more important than tillage practices (Gaynor et al. 1995; Sauer and Daniel 1987). Even under the worst-case scenario of intense rainfall within 1 wk of application, atrazine loss from surface runoff is less than 9% of the total amount applied. A small amount of rain enhances the movement of atrazine into the soil and thus considerably reduces atrazine runoff loss at later rainfall events (Gaynor et al. 1995). Runoff loss can be reduced when atrazine is incorporated into the soil (Baker and Laflen 1979; Hall et al. 1983). However, little reduction in runoff loss resulted when atrazine was applied beneath crop residues because atrazine is weakly adsorbed to crop residues (Baker et al. 1982). Atrazine runoff loss can be further reduced in well-drained soils (Southwick et al. 1990b).

Thus, over 90% of applied atrazine is retained in the soil, is subject to leaching loss, and is a potential threat for groundwater contamination. Atrazine movement in soils occurs mainly in the water phase and is affected by adsorption–desorption processes (Gaynor et al. 1995; Wienhold and Gish 1994). Many methods have been used to quantify atrazine movement in soils, such as thin-layered soil plates (Hubbs and Lavy 1990; Sanchez-Martin et al. 1994; Somasundaram et al. 1991), packed soil columns (Bouchard 1987; Elrick et al. 1966; Ma and Selim 1994b), intact or undisturbed soil columns (Edwards et al. 1992a; Gaber et al. 1995; Schiavon 1988a,b; Stehouwer et al. 1994), field tile-drains (Buhler et al. 1993; Johnson et al. 1995; Southwick et al. 1992), and lysimeters (Adams and Thurman 1991; Bowman 1990; Johnson et al. 1995; Kordel et al. 1992). Atrazine breakthrough curves (BTCs) from laboratory column studies show steep fronts and extensive tailing (Elrik et al. 1966; Gaber et al. 1995; Gamerdinger et al. 1991; Ma and Selim 1994b; Rao et al. 1979), especially at high water fluxes and for large aggregate sizes. Figure 8 provides an example of the effect of flow velocity on atrazine breakthrough results. This tailing behavior causes a long-lasting residue effect in the effluent or leaching water. Buhler et al. (1993) detected an average atrazine concentration of 0.29 $\mu$g/L in drainage water samples 3 yr after atrazine application and 93–140 $\mu$g/kg of extractable atrazine in the upper 15 cm of soil 26 mon after atrazine application. Such a slow release of atrazine from soils was primarily the result of adsorption–desorption hysteresis (Ma and Selim 1994b; Rao et al. 1979; Swanson and Dutt 1973).

Atrazine movement under field conditions is much less than that in soil columns when the same amount of water is applied because evaporation of water from the soil surface reduces downward movement and enhances upward movement of capillary water in the field (Hubbs and Lavy 1990; Leistra 1980). The increased adsorption on the soil surface caused by daily wetting and drying may also retard atrazine movement (White 1976). Starr and Glotfelty (1990) found that a large proportion of atrazine was on the surface soil after leaching with 10 cm of chloride solution. Von Stryk and

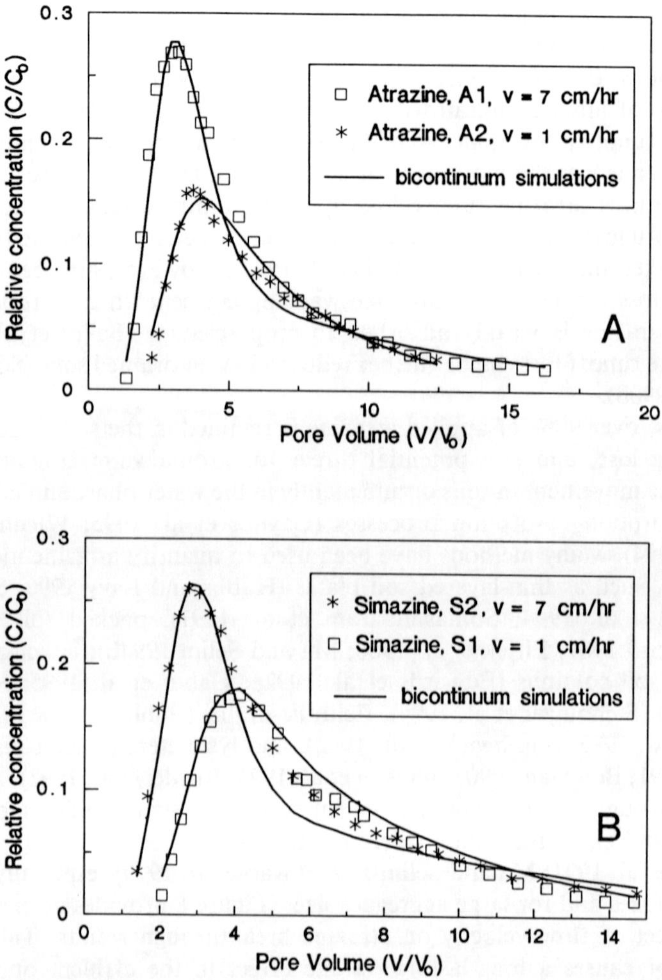


Fig. 8. Breakthrough curve of atrazine (A) and simazine (B) at two velocities on Valois soil; curves are fitted with a two-site model (bicontinuum model) (after Gamerdinger et al. 1991).

Bolton (1977) monitored atrazine from tile drains at a depth of 0.7 m in a clay soil and found that the total amount of atrazine leached per year was about 0.8% or less of the annual dosage. Muir and Baker (1978) estimated the loss of atrazine beyond 1.2 m from April to December to be about 0.22% of the annual dosage in a loamy soil. Southwick et al. (1992) obtained a total leaching loss of 0.6%–2.6% from 1-m-deep tile drains in a Sharkey clay soil 3 mon after application.

On the other hand, small amounts of atrazine can reach a greater depth than predicted because of preferential flow through cracks or large pores.

This preferential flow carries pesticides to deeper layers before the top layers are completely moistened (Edwards et al. 1993; Gish et al. 1991; Kladivko et al. 1991; Leistra 1980; Smith et al. 1992). Preferential flow of atrazine in soils with obvious macropores has been reported in both laboratory (Edwards et al. 1992a, 1993) and field experiments (Johnson et al. 1995; Kladivko et al. 1991). By comparing the time of rainfall and the time at which atrazine was detected in subsurface drains at a 1-m depth, Johnson et al. (1995) found that the breakthrough time for atrazine was much shorter than that calculated from saturated hydraulic conductivity, which suggests that atrazine bypassed the soil matrix and entered the drainage directly. Similar results were reported by Kladivko et al. (1991), who observed similar travel times for pesticides with a wide range of $K_{oc}$ values. The initiation of preferential flow depends on initial soil water content, organic matter content, rainfall intensity, timing of rainfall, macroporosity, and continuity of macropores (Edwards et al. 1992a, 1993; Shipitalo et al. 1990). Retarded atrazine migration through macropores observed in earthworm burrows was probably caused by organic linings on the surfaces of the burrow walls (Edwards et al. 1992b; Stehouwer et al. 1993, 1994).

Atrazine movement in the field is affected more by rainfall pattern or irrigation regimen than by agricultural practices (Edwards et al. 1993; Johnson et al. 1995; Sadeghi and Isensee 1992; Weed et al. 1995). In a field study, Troiano et al. (1993) found that atrazine movement was affected by both irrigation methods (sprinkler, basin, and furrow) and percolation treatments (low, medium, and high). Atrazine was leached to a deeper soil profile under a higher level of percolation rate with furrow irrigation. They also observed a prominent second atrazine peak lower in the soil profile. Johnson et al. (1995) found that atrazine concentration in tile drains was dependent on the timing and intensity of the first rainfall event. Immediate rain after atrazine application enhanced atrazine movement into drains. They also observed that a small rain after atrazine application enhanced atrazine distribution into the soil matrix and therefore reduced atrazine discharge into groundwater. Similar results have been reported by Edwards et al. (1993), Sadeghi and Isensee (1992), Schreiber et al. (1993), and Shipitalo et al. (1990). Edwards et al. (1993) applied a simulated storm onto undisturbed soil blocks either 1 h, 1 d, 1 wk, or 2 wk after atrazine application, and found that atrazine transport through earthworm burrows decreased with delay of the first rainfall. Therefore, preferential transport of atrazine is possible only if rainfall is of high intensity and occurs shortly after atrazine application. Wienhold and Gish (1994) also found that the timing of precipitation events affected atrazine volatilization under different tillage practices.

To minimize atrazine contamination of groundwater, controlled-release formulations of atrazine were tested. The most commonly used pesticide formulations are starch-encapsulation (Gish et al. 1994; Schreiber et al. 1993; Wienhold and Gish 1992), alginated encapsulation (Johnson and Pepperman 1995), and lignin entrapment (Riggle and Penner 1987). These formulations showed an extensive period of release time without greatly reducing pest control (Fleming et al. 1992; Schreiber et al. 1993). Formulated atrazine is released from the granules primarily by diffusion, which is controlled by water content, temperature, microbial activity, and specific surface area (Schreiber et al. 1993; Wienhold and Gish 1992). Atrazine granules can remain on the surface soil for an extended period of time under dry, cool, and nonfertile soil conditions. The half-life of atrazine in surface soils was 110 d for a starch-encapsulated formulation and 36 d for a commercial formulation (Gish et al. 1994). More atrazine was detected in the soil matrix when the starch-encapsulated formulation was used, which considerably reduced convective transport and preferential flow of atrazine (Schreiber et al. 1993). Encapsulated atrazine application also reduced volatilization loss (Wienhold and Gish 1994).

## VI. Modeling Approaches

Two types of models have been used in atrazine transport modeling: the stochastic approach (Utermann et al. 1990), and the deterministic approach based on detailed chemical and physical processes in the soil (Ma and Selim 1994b). The convective-dispersive equation (CDE) is commonly used in deterministic modeling of solute transport in soils, which can be written as

$$\frac{\partial C}{\partial t} = D\frac{\partial^2 C}{\partial x^2} - v\frac{\partial C}{\partial x} - \left[\frac{\rho}{\theta}\right]\frac{\partial S}{\partial t}, \tag{6}$$

where $\theta$ is the soil water content ($cm^3/cm^3$), $D$ is the hydrodynamic dispersion coefficient ($cm^2/hr$), $v$ is the average pore water velocity (cm/hr), $\rho$ is the soil bulk density ($g/cm^3$), $x$ is the spatial coordinate (cm), and $t$ is time (hr). The average pore water velocity is obtained from Darcy's water flux density ($q$) such that $v = q/\theta$.

### A. One-Adsorption-Site Models

The application and modification of the foregoing CDE [Eq. (6)] are mainly on the sorption term ($\partial S/\partial t$). The earliest approach to modeling pesticide movement in laboratory soil columns was the local equilibrium assumption (Elrick et al. 1966; Gaber et al. 1995; van Genuchten et al. 1974), where a linear adsorption isotherm, $S = K_d C$, was used. Therefore, a retardation factor $R$ ($= 1 + \rho K_d/\theta$) was introduced, and the CDE is thus written as

$$R\frac{\partial C}{\partial t} = D\frac{\partial^2 C}{\partial x^2} - v\frac{\partial C}{\partial x}. \tag{7}$$

This approach commonly provided a poor description of pesticide transport in soil columns (Davidson et al. 1968; Elrick et al. 1966; van Genuchten et al. 1974). Figure 9 shows the prediction of atrazine BTCs (dashed curves) using a $K_d$ value from batch experiments. Such an equilibrium model overpredicted the retardation and underpredicted the tailing of atrazine in soils. Swanson and Dutt (1973) used the Freundlich isotherm [Eq. (1)] and found that the predictions of atrazine BTCs were improved using different $N$ values ($N_a = 2.3\ N_d$) for adsorption and desorption processes. However, because atrazine adsorption is kinetic in nature, it is necessary to incorporate the rate of reaction of atrazine into transport models. The earliest kinetic approach is the linear first-order reaction, which may be expressed as (Travis and Etnier 1981; van Genuchten et al. 1974)

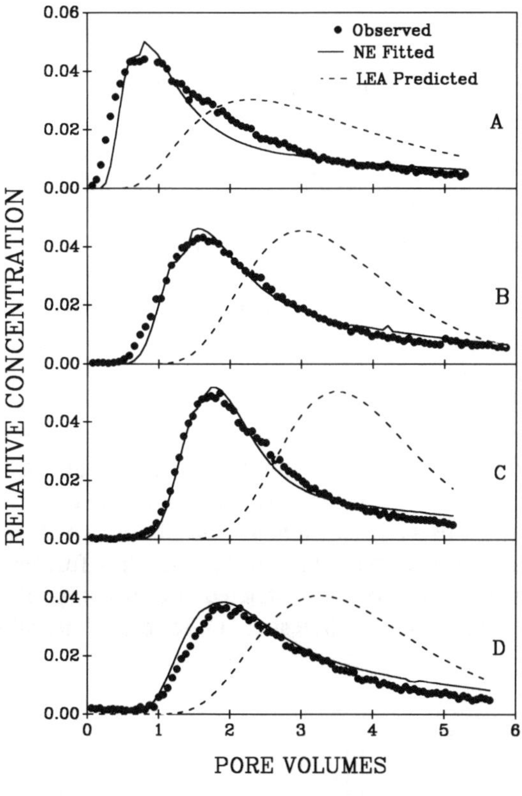


Fig. 9. Observed (circles) atrazine breakthrough curves (BCTs) in undisturbed soil columns (Amsterdam silt loam), fitted using the two-site nonequilibrium (NE) model (solid line), and predicted (dashed line) using the local equilibrium assumption: (A) fast, (B) medium (volumetric soil water content = 0.44), (C) medium, and (D) slow pore water velocity (after Gaber et al. 1995).

$$\frac{dS}{dt} = \kappa_1 C - \kappa_2 S, \tag{8}$$

where $\kappa_1$ and $\kappa_2$ are the forward and backward rate coefficients ($hr^{-1}$), respectively. A nonlinear first-order approach was later proposed (van Genuchten et al. 1974):

$$\frac{\partial S}{\partial t} = \kappa_1 C^N - \kappa_2 S, \tag{9}$$

where $N$ is reaction order with respect to $C$. This nonlinear model reduces to the Freundlich equation at equilibrium. However, the kinetic first-order models provided little success in predicting pesticide transport under different experimental conditions, especially for large soil aggregates and high water-flow velocity (Pignatello et al. 1993; van Genuchten et al. 1974).

## B. Two-Adsorption-Site Models

The one-adsorption-site approaches were based on the assumption that the soil is chemically homogeneous, which may not be true for most soil–pesticide combinations. Some soil components have a higher affinity to atrazine than others. This resulted in the proposal of multireaction-site models (Selim et al. 1976, 1992). One of the simplest chemical nonequilibrium models is the two-adsorption-site models of Selim et al. (1976). The two-site model improved the description of atrazine in soils (Gaber et al. 1995; Gamerdinger et al. 1991; Pignatello et al. 1993; Rao et al. 1979; Wauchope and Myers 1985). Figures 8 and 9 show two applications of the two-site model on atrazine transport.

The two-site model was also supported experimentally. Clay and Koskinen (1990a,b) found that 98% of atrazine that was adsorbed within 24 hr was adsorbed in the first 2 hr and that the change in the amount of atrazine adsorbed between 24 and 48 hr was less than 1%–2 %. Wauchope and Myers (1985) also observed two adsorption phases which were denoted as labile and restricted atrazine absorption sites. They further showed an increase in atrazine adsorption on restricted sites accompanied by a concurrent decrease of atrazine on labile sites. Other experimental support of the

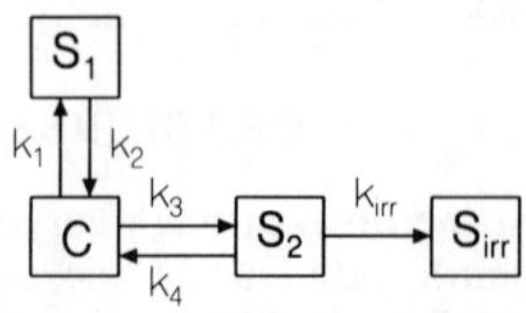


Fig. 10. A schematic of the proposed two-retention-site model with a first-order degradation reaction. $S_1$, $S_2$, amounts of atrazine sorbed on types 1 and 2 reaction sites, respectively; $S_{irr}$, amount irreversibly retained; $C$, soil solution phase.

two-site model is provided by the quite different affinity of organic matter and clay minerals to atrazine (Wang et al. 1992).

Ma and Selim (1994b) applied the two-adsorption site model to atrazine retention and transport in a Sharkey clay soil, where a first-order degradation of kinetically adsorbed atrazine was assumed because atrazine degradation is soil surface catalyzed (Fig. 10). Mathematical formulation of the two-site model is

$$\frac{\partial S_1}{\partial t} = \frac{\theta}{\rho} \kappa_1 C^M - \kappa_2 S_1, \tag{10}$$

$$\frac{\partial S_2}{\partial t} = \frac{\theta}{\rho} \kappa_3 C^N - (\kappa_4 + \kappa_{irr}) S_2, \tag{11}$$

and a first-order degradation reaction can be written as

$$\frac{\partial S_{irr}}{\partial t} = \kappa_{irr} S_2, \tag{12}$$

where $S_1$ and $S_2$ are the amounts of atrazine sorbed on type 1 and type 2 reaction sites ($\mu$g/g), respectively, $S_{irr}$ is the amount irreversibly retained ($\mu$g/g), $\kappa_1$ and $\kappa_3$ are the forward reaction rate coefficients associated with $S_1$ and $S_2$ (hr$^{-1}$), respectively, and $\kappa_2$ and $\kappa_4$ are the corresponding backward rate coefficients associated with $S_1$ and $S_2$ (hr$^{-1}$), respectively. In addition, $\kappa_{irr}$ is the rate coefficient associated with the irreversible phase $S_{irr}$. If $\kappa_1$ and $\kappa_2$ are very large, Eq. (10) can be viewed as an equilibrium reaction and be rewritten as

$$S_1 = \frac{\theta}{\rho} \frac{\kappa_1}{\kappa_2} C^M = \frac{\theta}{\rho} K_e C^M, \tag{13}$$

where $K_e$ ($= \kappa_1/\kappa_2$) is a dimensionless equilibrium parameter. Thus, the sorption term ($\partial S/\partial t$) for the equilibrium/kinetic two-site model can be implicitly expressed as

$$\frac{\partial S}{\partial t} = K_e M C^{M-1} \frac{\theta}{\rho} \frac{\partial C}{\partial t} + \frac{\theta}{\rho} \kappa_3 C^N - \kappa_4 S_2. \tag{14}$$

Usually, $M$ and $N$ are assumed to be equal and are estimated from batch isotherms using the Freundlich equation (Ma and Selim 1994b; Selim et al. 1976). Thus, the differences between the two sites result only from their reaction rate coefficients. Ma and Selim (1994b) found that the two-site model provided a good description for both batch and column transport data, but its predictability was very poor (Figs. 11 and 12).

Pignatello et al. (1993) formulated a two-adsorption-site model based on diffusion controlled processes, in which $S_1$ was assumed to be in equilibrium with the solution phase and $S_2$ was controlled by diffusion. They found that their mathematical formula described atrazine transport better than that of Selim et al. (1976).

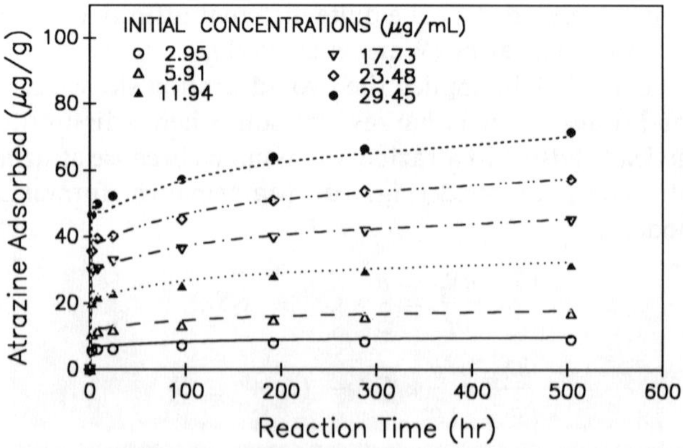


Fig. 11. Atrazine adsorption at soil : water ratio of 1 : 6 versus time in a Sharkey soil. Curves are fitted with the multireaction model to all the initial concentrations simultaneously (after Ma and Selim 1994b).

## C. Second-Order Two-Site Models

One of the assumptions associated with the two-site model is that the adsorption sites in the soil are unlimited, which may not be true for atrazine (Gamble and Khan 1990, 1992; Gilchrist et al. 1993; Wang et al. 1992). Selim and Amacher (1988) proposed a second-order two-site model in

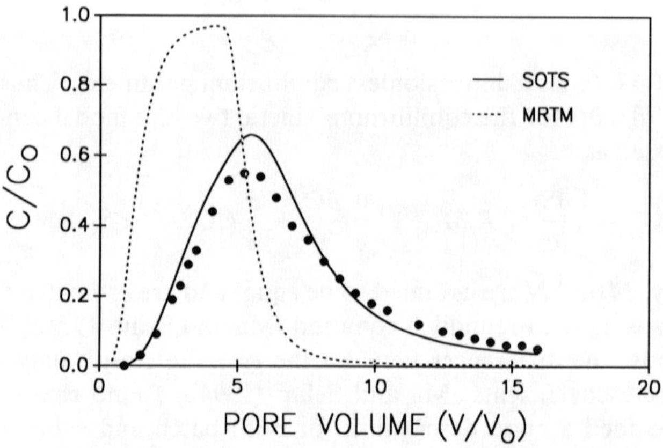


Fig. 12. Measured (*solid circles*) and predicted (*lines*) atrazine BCTs with 2–4 mm aggregated Sharkey soil, column length of 15 cm, velocity of 1.08 cm/hr, and initial concentration of 10.15 μg/mL. $C_o$ is the concentration of the input solution. Atrazine predictions were based on batch-derived model parameters. SOTS; Second-order two-site model. MRTM; Multireaction transport model (two-adsorption-site model) (after Ma and Selim 1994b).

which the total adsorption sites on the soil are limited and are partitioned between $S_1$ and $S_2$. The reaction rate is thus a function not only of solute present in solution but also the amount of available (or vacant) sites on soil surfaces. It is also assumed that the total retention capacity or the total amount of sites on soil surfaces $S_T$ ($\mu$g/g soil) is time-invariant and can be estimated from adsorption isotherms using the Langmuir equation [Eq. (2)].

Ma and Selim (1994a) proposed a modified second-order formulation in which they assumed that vacant sites are equally accessible to $S_1$ and $S_2$. That is, $S_1$ and $S_2$ can compete for the unoccupied adsorption sites regardless of whether they are of type 1 or type 2 sites. In addition, because atrazine and hydroxyatrazine occupied distinct adsorption sites in the soil, $S_{irr}$ (assumed to be mainly hydroxyatrazine) was excluded from the total adsorption sites. Therefore, the total adsorption sites ($S_T$) can be defined as

$$S_T = \phi + S_1 + S_2, \qquad (15)$$

where $\phi$ is the total vacant sites in the soil. The governing kinetic reaction rates for atrazine present in the soil solution phase ($C$) and that in the reversible and irreversible phases ($S_1$, $S_2$, and $S_{irr}$) may be written as

$$\frac{\partial S_1}{\partial t} = k_1\theta C\phi - k_2 S_1 \qquad (16)$$

$$\frac{\partial S_2}{\partial t} = k_3\theta C\phi - (k_4 + k_{irr})S_2 \qquad (17)$$

$$\frac{\partial S_{irr}}{\partial t} = k_{irr}S_2, \qquad (18)$$

where $k_1$ and $k_3$ are forward rate coefficients (mL $\mu g^{-1}$ $hr^{-1}$), and $k_2$ and $k_4$ are the corresponding backward rate coefficients ($hr^{-1}$) associated with the second-order model. Assuming that $S_1$ is in equilibrium, the adsorption term $\partial S/\partial t$ for a modified second-order equilibrium/kinetic model can be expressed as

$$\frac{\partial S}{\partial t} = K\theta\phi\frac{\partial C}{\partial t} + k_3\theta C\phi - k_4 S_2, \qquad (19)$$

where $K$ ($= k_1/k_2$) is an equilibrium parameter associated with $S_1$ (mL/$\mu$g). The CDE can be written as

$$(1 + K\rho\phi)\frac{\partial C}{\partial t} = D\frac{\partial^2 C}{\partial x^2} - v\frac{\partial C}{\partial x} + \left[\frac{\rho}{\theta}\right](k_3\theta C\phi - k_4 S_2). \qquad (20)$$

Ma and Selim (1994a) found that this model not only described atrazine kinetics at several initial concentrations but also successfully predicted atrazine adsorption–desorption under different soil : water ratios and desorption hysteresis. Using model parameters derived from batch experiments, Ma and Selim (1994b) were able to predict atrazine BTCs from soil columns under different experimental conditions, such as velocity, column length, aggregate size, and input concentration (Fig. 12).

## D. Original Mobile–Immobile Two-Region Models

The predictability of chemical nonequilibrium models often decreased with increasing soil aggregate size and water flow velocity (Ma and Selim 1994b; van Genuchten et al. 1974). However, improved predictions of pesticide movement at high water fluxes were obtained by assuming that only a fraction of the soil participates in adsorption processes (van Genuchten et al. 1974). This finding supported the proposal of physical nonequilibrium models (Skopp and Warrick 1974; van Genuchten and Wierenga 1976). The most commonly used physical nonequilibrium model is the mobile–immobile approach, which is mathematically equivalent to the models based on chemical nonequilibrium (two-site model) under certain assumptions, for example, linear adsorption (Nkedi-Kizza et al. 1984; van Genuchten and Wagenet 1989). The mobile–immobile approach has been applied to atrazine modeling in recent literature (Gaber et al. 1995; Selim and Ma 1995; Rao et al. 1979). In the mobile–immobile approach, soil water is divided into mobile ($\theta^m$) and immobile ($\theta^{im}$) phases based on its flow velocity (Fig. 13). The mobile phase contributes to water flow in the effluent, and the immobile phase does not contribute to flow. The soil was also divided into two regions; one is in direct contact with the mobile water and is referred to as the dynamic soil region, and the other is in direct contact with the immobile water and is thus called the stagnant region (van Genuchten and Wierenga 1976). Furthermore, the dynamic soil region can only react with solutes in the mobile phase, while the stagnant soil region can only react with solutes in the immobile phase. The CDE for reactive solutes in the mobile–immobile model can be written as

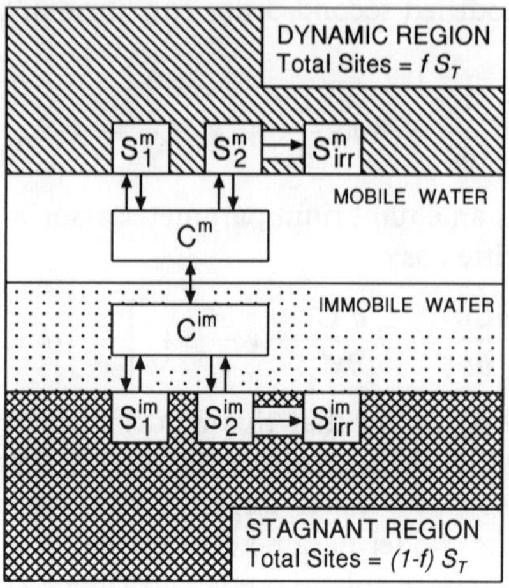


Fig. 13. A schematic of the original two-region model (after Selim and Ma 1995).

$$\theta^{\mathrm{m}} \frac{\partial C^{\mathrm{m}}}{\partial t} + f\rho \frac{\partial S^{\mathrm{m}}}{\partial t} = \theta^{\mathrm{m}} D \frac{\partial^2 C^{\mathrm{m}}}{\partial x^2} - \nu^{\mathrm{m}} \theta^{\mathrm{m}} \frac{\partial C^{\mathrm{m}}}{\partial x} - \alpha(C^{\mathrm{m}} - C^{\mathrm{im}}), \quad (21)$$

where $C^{\mathrm{m}}$ and $C^{\mathrm{im}}$ are solute concentrations in the mobile and immobile phases ($\mu$g/mL); and $S^{\mathrm{m}}$ and $S^{\mathrm{im}}$ are the amounts of solute sorbed in the dynamic and stagnant soil regions ($\mu$g/g), respectively. $\nu^{\mathrm{m}}$ $(=q/\theta^{\mathrm{m}})$ is the average pore water flux of the mobile phase (cm/hr), and $q$ is the water flux density (cm/hr). The parameter $f$ is the fraction of the dynamic soil region or the fraction of adsorption sites available to the mobile water phase if the adsorption sites are uniformly distributed on the soil matrix. The mass-transfer equation governing the interaction between the mobile and immobile phases is

$$\theta^{\mathrm{im}} \frac{\partial C^{\mathrm{im}}}{\partial t} + (1 - f)\rho \frac{\partial S^{\mathrm{im}}}{\partial t} = \alpha(C^{\mathrm{m}} - C^{\mathrm{im}}), \quad (22)$$

where $\alpha$ is the mass-transfer coefficient ($\mathrm{hr}^{-1}$). This model was capable of explaining preferential flow and tailing of BTCs (Rao et al. 1979; van Genuchten et al. 1977). Chemical reactions in each region are usually assumed to be identical to avoid the added difficulties in estimation of parameters associated with each region. The simplest chemical reaction introduced into the mobile-immobile approach is the equilibrium type (Gaber et al. 1995; Nkedi-kizza et al. 1984; Rao et al. 1979; van Genuchten and Wierenga 1976). However, the equilibrium assumption usually did not provide adequate predictions of pesticide transport when model parameters were estimated under different experimental conditions (Rao et al. 1979). Later, kinetic reactions were introduced into the mobile–immobile model (Brusseau et al. 1989; van Genuchten and Wagenet 1989). Selim and Amacher (1988) combined the mobile–immobile model with a second-order, two-site chemical nonequilibrium approach. Because it is difficult to determine the parameter $f$ of the mobile–immobile model, it is often difficult to carry out comprehensive model validation. Thus, the model can only be used to describe or simply "fit" experimental data in question (Gaber et al. 1995; Rao et al. 1979; van Genuchten and Wierenga 1976). On the other hand, the parameter $f$ must be defined for the purpose of model predictions. Two assumptions of $f$ have been reported: one is $f = 0$, where solute has to diffuse to the immobile water to react with adsorption sites, and the other assumes that $f$ equals the fraction of mobile water (Nkedi-kizza et al. 1983). Selim and Ma (1995) found that both assumptions were incapable of predicting atrazine BTCs (Fig. 14).

## E. Modified Mobile–Immobile Two-Region Models

Selim and Ma (1995) proposed a new modified conceptual mobile-immobile model in which the dynamic and stagnant soil regions are regarded as a continuum and connected to one another. Because a soil aggregate may be surrounded by both mobile and immobile water phases, as indicated by van Genuchten and Wierenga (1976), retention reactions occur concurrently in

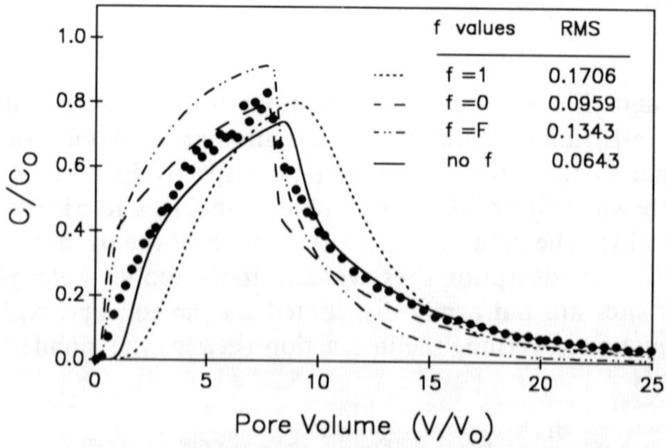


Fig. 14. Measured (*solid circles*) and predicted (*lines*) atrazine BCTs with 4–6 mm aggregated Sharkey soil, column length of 10 cm, velocity of 1.76 cm/hr, and initial concentration of 11.33 μg/mL. $C_o$ is the concentration of the input solution. Atrazine predictions were based on mobile–immobile approaches, and the model parameters were from batch kinetic experiments. *F* is the fraction of mobile water (after Selim and Ma 1995).

both the dynamic and stagnant regions until equilibrium is attained inside the soil aggregates (Fig. 15). Conceptually, the new model assumed that all available or vacant sites on the soil matrix are potentially accessible to the solute present in both the mobile and immobile water phases. Thus, solute adsorption from each water phase is a function of the total vacant sites in the soil. Therefore, there is no distinct delineation between the sites associated with the two soil regions. As a result, the parameter *f* is not necessary

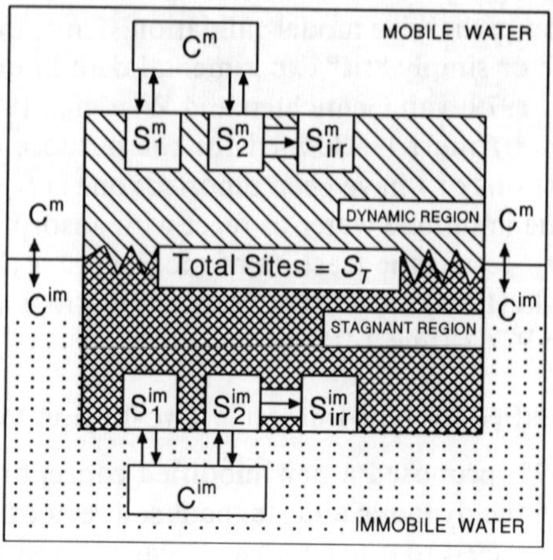


Fig. 15. A schematic of the modified two-region model (after Selim and Ma 1995).

in the new two-region approach, and in fact $f$ was found to be highly affected by experimental conditions, such as particle size, water flux, solute concentration, and solute type (Nkedi-Kizza et al. 1983; van Genuchten and Wierenga 1977). Based on these concepts, the modified CDE without $f$ can be written as (Selim and Ma 1995)

$$\theta^{\mathrm{m}} \frac{\partial C^{\mathrm{m}}}{\partial t} + \rho \frac{\partial S^{\mathrm{m}}}{\partial t} = \theta^{\mathrm{m}} D \frac{\partial^2 C^{\mathrm{m}}}{\partial x^2} - \nu^{\mathrm{m}} \theta^{\mathrm{m}} \frac{\partial C^{\mathrm{m}}}{\partial x} - \alpha(C^{\mathrm{m}} - C^{\mathrm{im}}), \quad (23)$$

and the associated mass-transfer equation is

$$\theta^{\mathrm{im}} \frac{\partial C^{\mathrm{im}}}{\partial t} + \rho \frac{\partial S^{\mathrm{im}}}{\partial t} = \alpha(C^{\mathrm{m}} - C^{\mathrm{im}}). \quad (24)$$

Selim and Ma (1995) incorporated the modified second-order equilibrium/kinetic model into the new mobile–immobile approach. Let $\phi$ denote the total amount of vacant sites. The reactions in the dynamic soil region are

$$S_1^{\mathrm{m}} = K\theta^{\mathrm{m}} C^{\mathrm{m}} \phi, \quad (25)$$

$$\frac{\partial S_2^{\mathrm{m}}}{\partial t} = k_3 \theta^{\mathrm{m}} C^{\mathrm{m}} \phi - (k_4 + k_{\mathrm{irr}}) S_2^{\mathrm{m}}, \quad (26)$$

$$\frac{\partial S_{\mathrm{irr}}^{\mathrm{m}}}{\partial t} = k_{\mathrm{irr}} S_2^{\mathrm{m}}. \quad (27)$$

For the stagnant region, the reactions are

$$S_1^{\mathrm{im}} = K\theta^{\mathrm{im}} C^{\mathrm{im}} \phi, \quad (28)$$

$$\frac{\partial S_2^{\mathrm{im}}}{\partial t} = k_3 \theta^{\mathrm{im}} C^{\mathrm{im}} \phi - (k_4 + k_{\mathrm{irr}}) S_2^{\mathrm{im}}, \quad (29)$$

$$\frac{\partial S_{\mathrm{irr}}^{\mathrm{im}}}{\partial t} = k_{\mathrm{irr}} S_2^{\mathrm{im}}, \quad (30)$$

where the sorption capacity $S_{\mathrm{T}}$ is related to the total vacant sites $\phi$, according to

$$S_T = \phi + S_1^{\mathrm{m}} + S_2^{\mathrm{m}} + S_1^{\mathrm{im}} + S_2^{\mathrm{im}}. \quad (31)$$

A major advantage of this new model is that the model can be used to predict atrazine BTCs using batch-estimated reaction coefficients. The amount of atrazine adsorbed in each soil region is not adjusted by $f$ but is solely determined by reaction rates. Selim and Ma (1995) tested this model versus the original mobile–immobile model with $f = 1$, $f = 0$, and $f = F$ (the fraction of mobile water). They found that the new mobile–immobile approach provided better predictions of atrazine breakthrough results under a wide range of experimental conditions (flow velocity, flow interruption, aggregate size, initial concentration, and column length) over the original mobile–immobile model (Fig. 14).

## F. Transport Models at the Field Scale

Atrazine transport is much more complicated under field conditions than under laboratory conditions. Factors contributing to the difficulty of measuring and modeling atrazine transport under field conditions include soil spatial heterogeneity and irregular weather sequence. Experimentally, atrazine transport in soils is commonly measured by periodically monitoring its presence in the soil (taking soil cores) and water solution (from subsurface drains and suction lysimeters) after application. Such measurements are highly dependent on sampling methods and rarely represent atrazine behavior at the field scale due to spatial and temporal variability (Clay et al. 1994). Such an experimental difficulty results in unsuccessful applications of models at the field scale (Pennell et al. 1990). As a result, criteria for model calibration and validation at the field scale cannot be as rigid as that for laboratory data. Several criteria for evaluating field-oriented models are peak concentration, peak position, maximum leaching depth, travel time to certain depth, and accumulative leachate.

Several field-oriented models have been documented, such as the Chemical Movement in Layer Soils (CMLS) (Nofziger and Hornsby 1986), Method of Underground Solute Evaluation (MOUSE) (Steenhuis et al. 1987), Pesticide Root Zone Model (PRZM) (Carsel et al. 1985), Groundwater Loading Effects of Agricultural Management Systems (GLEAMS) (Leonard et al. 1987), Leaching Estimation and Chemistry Model-Pesticides (LEACHMP) (Hutson and Wagenet 1993; Wagenet and Hutson 1986), and Root Zone Water Quality Model (RZWQM) (Ahuja et al. 1993). All models assume homogeneous soil structure (or layered soil) and equilibrium adsorption, although some models are capable of incorporating a first-order kinetic reaction and first-order degradation. Sauer et al. (1990) evaluated PRZM on atrazine mobility under two tillage systems, and they found that PRZM yielded good simulation in the early growing season, including peak position and maximum penetration depth in the soil profile, while it overpredicted pesticide penetration at later sampling dates. In addition, PRZM simulation overpredicted peak concentration in early sampling dates and underpredicted peak concentration in later sampling dates. This is expected on the basis of the equilibrium adsorption assumption. Other examples of model applications with varied degree of success include those of Thooko et al. (1994) and Sichani et al. (1991) with GLEAMS and Smith et al. (1991) with PRZM and GLEAMS.

Because it is difficult to obtain well-defined sets of experimental data for model verification, some models can be used only to simulate the effect of environmental factors on possible atrazine behavior. Ahuja et al. (1993) studied the effect of macroporosity, antecedent water content, rainfall intensity, evaporation, and transpiration on atrazine distribution in soils using RZWQM. Haan et al. (1994) coupled the CMLS with a random weather generator (WGEN) to study the travel time probability of atrazine under different weather sequences. They found that the travel time to reach 0.5 m can range from 200 to 4000 d depending on weather conditions.

On the other hand, Hutson and Wagenet (1993) simulated accumulative atrazine leachate at a 1.1-m depth using LEACHMP with 10-yr daily recorded rainfall and temperature data from Aurora, New York. Comparing the simulated results with soil cores taken from the field, they found that LEACHMP overpredicted the penetration depth of atrazine. Because mathematical models cannot elaborate all processes and usually contain several unknown model parameters, even for simplified systems, such models are not satisfactorily evaluated. Thus, simple empirical models may be more useful than those based on several detailed chemical and physical processes (Nicholls et al. 1982).

### G. Model Parameter Estimations

All models are conceptual in nature. Therefore, their application to real soil systems can be difficult and sometimes have great uncertainty. With the increasing degree of complexity, more and more model parameters are needed. Model parameters are usually introduced without proposing direct or indirect experimental methods for their measurements. Thus, model parameters are optimized by minimizing the sum of square errors (van Genuchten 1981) or the largest percentage difference between the model simulation and experimental observation (Pignatello et al. 1993). The optimized parameters are not truly constant and depend on experimental conditions (Gaber et al. 1995; Gamerdinger et al. 1991). One set of experimental data can often be fitted equally well with several model formulations (Ma and Selim 1994a,b). Such an empirical modeling approach increases the difficulty in parameter estimation.

Batch experiment is commonly used to derive model parameters related to chemical reactions, such as $K_d$. Although this approach may not yield suitable model parameters for atrazine transport, it can be used to compare atrazine with other pesticides in different soils. $K_d$ is usually obtained by fitting the adsorption isotherm to a linear equation. It may also be estimated from $K_{oc}$ (Green and Karickhoff 1990; Singh et al. 1989). The rate coefficients ($k_i$) for chemical nonequilibrium models can be derived from batch kinetic experiments or column miscible displacement experiments (Ma and Selim 1994a,b). Fitted rate coefficients are usually dependent on flow velocity and initial atrazine concentration (Gaber et al. 1995; Gamerdinger et al. 1991; Gaston and Locke 1994), and this is expected for models with empirical rate coefficients. To obtain concentration-independent model parameters, Ma and Selim (1994a) fitted their model to atrazine kinetic adsorption data for several initial concentrations simultaneously. Gamble and co-workers introduced an on-line microfiltration and HPLC analysis technique to speciate atrazine adsorption components and associated model parameters (Gamble and Khan 1990, 1992; Gilchrist et al. 1993).

In a study with the herbicide alachlor [2-chloro-*N*-(2,6-diethylphenyl)-*N*-(methoxymethyl) acetamide], Gaston and Locke (1994) related the conceptual adsorption sites ($S_1$, $S_2$, and $S_{irr}$) to the amount of alachlor extracted with different solvents. They found that the model parameters estimated from both solution concentrations ($C$) and estimated adsorptions ($S_1$, $S_2$, and $S_{irr}$) provided better descriptions of alachlor than using solution concentration ($C$) alone. Parameters for field-oriented models are more difficult to estimate because there is considerable uncertainty in environmental factors, however. Since physical nonequilibrium is perhaps dominant under field conditions, it is not appropriate to fit field observations to chemical nonequilibrium models without full understanding of the physical processes. In addition, field data are often scattered or poorly defined and cannot be used to calibrate models. Therefore, model parameters for field models are usually estimated from laboratory experiments.

Physical nonequilibrium models are conceptual as well, and associated model parameters are not well defined and are often controversial, such as mobile water ($\theta^m$), mass transfer coefficient ($\alpha$), and fraction of sites ($f$) (Ma and Selim 1995; Selim and Ma 1995). $\theta^m$ is usually estimated from curve-fitting, water content at certain tension, or hydraulic conductivity. The $\alpha$ is commonly used as a fitting parameter, although there are two methods to estimate $\alpha$ based on diffusion models (Rao et al. 1980; van Genuchten and Dalton 1986). Clothier et al. (1992) and Jaynes et al. (1995) have proposed an experimental method to estimate $\theta^m$ and $\alpha$ from an infiltration study of nonreactive tracers. The parameter $f$ can only be obtained through curve-fitting. Due to the uncertainty of model parameters, few models can be applied in a prediction mode. Models with few unknown parameters are more desirable to validate a model. Ma and Selim (1994a,b) and Selim and Ma (1995) introduced modified models that require fewer model parameters, without loss of predictability. As a matter of fact, their modified models are perhaps an improvement over the original model because there is less uncertainty of model parameters. A complex mathematical model with overloaded parameters therefore may not necessarily represent the real soil system unless it can be experimentally validated.

## VII. Conclusions

Atrazine retention and transport have been investigated for some 40 years. Experimentally, most studies are on the effects of soil and environmental conditions on atrazine retention and transport, and they have contributed to a massive quantity of information on atrazine–soil interactions. Because of the complexity of the soil system, several experimental data sets are scattered and sometimes inconsistent. The mechanisms of atrazine–soil interactions are not fully understood and need to be investigated with novel methodologies. Batch and miscible displacement experiments are the commonly accepted techniques to study atrazine retention and transport in the laboratory, and the amount of atrazine adsorbed is usually calculated from the concentration change in the solution phase. The batch method may only provide the maximum adsorption because of the breakdown of soil aggregates and mixing of soil solution. Thus, information obtained from batch experiments can be of only limited value to transport.

Miscible displacement has been used to study atrazine adsorption kinetics and transport in the laboratory. However, the results from miscible displacement experiments are sometimes not transferable to field observations because of spatial variability. Field studies of atrazine are usually conducted by taking representative soil (surface soil or soil core) and water samples (suction lysimeter, surface runoff, or subsurface drainage) at certain scheduled times after application. The sampling time and sampling methods are always debatable as to their representability of field conditions. Soil and water samples are taken to the laboratory for atrazine extraction and analysis. Methods of extraction and analysis vary from laboratory to laboratory, and extraction efficiency depends on atrazine-soil reaction time as well as on the extraction solvent. The diversity of experimental methods in the literature (or sometimes inappropriate application) may prevent knowledge transfer from laboratory to laboratory or from laboratory to the field.

To understand the effect of experimental condition on atrazine retention and transport, a number of mathematical models with different degrees of complexity have been developed. Equilibrium models (adsorption isotherm) are usually used to describe batch results. The convective-dispersive equation coupled with various chemical reaction models is often applied to interpret miscible experimental results. Two types of transport models are proposed and tested in the laboratory: one is based on chemical nonequilibrium and the other on physical nonequilibrium. The latter may be more appropriate under field conditions where macropore flow often plays a significant role. Because all models are conceptual, the parameters associated with each model are also conceptual. Therefore, a difficulty in model development and application is how to obtain the necessary model parameters. Estimated model parameters may depend on experimental conditions, such as flow velocity, initial atrazine concentration, and soil structure. As a result, it is difficult to predict atrazine behavior using independently derived model parameters (model validation). This difficulty may be overcome either by developing new methods of estimating model parameters or by introducing fewer model parameters. Simple models with fewer parameters are more desirable for field application, where a unique set of parameters are often difficult to quantify.

Validation of models under different experimental conditions is an important step for their evaluation. This is often difficult to fully achieve, especially for field-oriented models, because of the lack of well-defined data sets as well as model incapability. If a model is capable of predicting atrazine retention and transport under a given experimental constraint, the model does not necessarily reflect the mechanisms governing atrazine behavior in soils. In fact, other model formulae may be equally capable of such predictions because all the mathematical models are empirically based.

## Summary

No pesticide has been studied as extensively as atrazine. The study of atrazine has contributed to our general understanding of the behavior of pesticides in soils. New knowledge and concepts were evaluated, such as atrazine adsorption kinetics, desorption hysteresis, and preferential flow. Corresponding conceptual models were also proposed to explain the behavior of atrazine in soils. Atrazine adsorption–desorption is the major process affecting atrazine behavior in soils and is mainly affected by organic matter and soil pH. Atrazine in soils is subject to biological and chemical degradations. Hydroxyatrazine, the chemical degradation product, is more strongly adsorbed to soil than atrazine. Deethylatrazine and deisopropylatrazine, the major biological degradation products, are more mobile than atrazine. Hydrolysis of atrazine is soil-surface catalyzed and favored by low soil pH. The overall dissipation rate of atrazine was found to be pseudo first-order.

Two distinct and different processes are involved in atrazine movement: slow transport through the soil matrix and rapid movement through macropores. The first process is controlled by adsorption kinetics and degradation reactions and can be well explained by models based on chemical heterogeneity, such as the two-site models and second-order models. The second flow process results from preferential flow through large pores and can be explained by physical nonequilibrium models such as the mobile–immobile and two-flow domain models. Because both processes coexist in atrazine transport, coupling of physical and chemical nonequilibrium models is often necessary and has shown promise in atrazine transport modeling. However, more efforts are needed in estimating model parameters and in developing management-oriented models.

## References

Adams CD, Thurman EM (1991) Formation and transport of deethylatrazine in the soil and vadose zone. J Environ Qual 20:540–547.

Ahuja LR, DeCoursey DG, Barnes BB, Rojas KW (1993) Characteristics of macropore transport studied with the ARS root zone water quality model. Trans ASAE 36:369–380.

Albanis TA, Pomonis PJ, Sdoukos AT (1989) The influence of fly ash on pesticide fate in the environment: II. Hydrolysis: degradation and adsorption of atrazine in aqueous mixtures of soil with fly ash. Toxicol Environ Chem 19:171–178.

Alhajjar BJ, Simsionan GV, Chesters G (1990) Fate and transport of alachlor, metolachlor and atrazine in large columns. Water Sci Technol 22:87–94.

Anonymous (1990) Iowa restricts atrazine use on certain soils. Success Farm 88:58.

Ashton FM, Monaco TJ (1991) Weed Science, Principles and Practices, 3rd Ed. John Wiley & Sons, New York.

Aspelin AL (1994) Pesticides industry sales and usage—1992 and 1993 market estimates. OPP, Office of Prevention, Pesticides and Toxic Substance, U.S. Environmental Protection Agency, Washington, DC. p 13.

Assaf NA, Turco RF (1994) Influence of carbon and nitrogen application on the mineralization of atrazine and its metabolites in soil. Pestic Sci 41:41–47.

Bacci E, Renzoni A, Gaggi C, Calamari D, Franchi A, Vighi M, Severi A (1989) Models, field studies, laboratory experiments: an integrated approach to evaluate the environmental fate of atrazine (*s*-triazines). Agric Ecosys Environ 27:513–522.

Baker JL, Laflen JM (1979) Runoff losses of surface-applied herbicides as affected by wheel tracks and incorporation. J Environ Qual 8:602–607.

Baker JL, Laflen JM, Hartwig RO (1982) Effects of corn residue and herbicide placement on herbicide runoff losses. Trans ASAE 25:340–343.

Baluch HU, Somasundaram L, Kanwar RS, Coats JR (1993) Fate of major degradation products of atrazine in Iowa soils. J Environ Sci Health B28:127–149.

Barriuso E, Schiavon M, Anderux F, Portal JM (1991) Location of atrazine non-extractable (bound) residues in soil size fractions. Chemosphere 22:1131–1140.

Barriuso E, Baer U, Calvet R (1992) Dissolved organic matter and adsorption-desorption of dimefuron, atrazine, and carbetamide by soils. J Environ Qual 21: 359–367.

Barriuso E, Laird DA, Koskinen WC, Dowdy RH (1994) Atrazine desorption from smectites. Soil Sci Soc Am J 58:1632–1638.

Basta NT, Olness A (1992) Determination of alachlor, atrazine and metribuzin in soil by resin extraction. J Environ Qual 21:497–502.

Behki RM, Khan SU (1986) Degradation of atrazine by *Pseudomonas*: *N*-dealkylation and dehalogenation of atrazine and its metabolites. J Agric Food Chem 34:746–749.

Behki RM, Khan SU (1994) Degradation of atrazine, propazine and simazine by *Rhodococcus* strain B-30. J Agric Food Chem 42:1237–1241.

Bertin G, Schiavon M, Pottier C (1990) Plant bioavailability of "natural" and "model" humic acid bound $^{14}C$ atrazine residues. Toxicol Environ Chem 26:203–210.

Best JA, Weber JB (1974) Disappearance of *s*-triazines as affected by soil pH using a balance sheet approach. Weed Sci 22:364–373.

Boesten JJTI (1990) Influence of solid/liquid ratio on the experimental error of sorption coefficients in pesticide/soil systems. Pestic Sci 30:31–41.

Bouchard DC (1987) Monitoring transport of selected pesticides and phenols in soil columns by high performance liquid chromatograph. J Environ Sci Health B22: 391–402.

Bowman BT, Sans WW (1985) Partitioning behavior of insecticides in soil-water systems. II. Desorption hysteresis effects. J Environ Qual 14:270–273.

Bowman BT (1990) Mobility and persistence of alachlor, atrazine and metolachlor in planifield sand and a triazine and isazofos in honeywood silt loam using field lysimeters. Environ Toxicol Chem 9:453–461.

Briggs GG (1981) Theoretical and experimental relationship between soil adsorption, octanol–water partition coefficients, water solubility, bioconcentration factors, and the parachlor. J Agric Food Chem 29:1050–1059.

Brouwer WWM, Boesten JJTI, Siegers WG (1990) Adsorption of transformation products of atrazine by soil. Weed Res 30:123–128.

Brusseau ML, Jessup RE, Rao PSC (1989) Modeling the transport of solutes influenced by multiprocess nonequilibrium. Water Resour Res 25:1971–1988.

Buhler DD, Randall GW, Koskinen WC, Wyse DL (1993) Atrazine and alachlor losses from subsurface tile drainage of a clay loam soil. J Environ Qual 22:583–588.

Burkart MR, Kolpin DW (1993) Hydrology and land-use factors associated with herbicides and nitrate in near-surface aquifers. J Environ Qual 22:646–656.

Burkhard N, Guth JA (1981) Chemical hydrolysis of 2-chloro-4,6 bis(alkylamino)-1,3,5-triazine herbicides and their breakdown in soil under influence of adsorption. Pestic Sci 12:45–52.

Calvet R (1980) Adsorption-desorption phenomena. In: Hance RJ (ed) Interaction Between Herbicides and the Soil. Academic Press, London, pp 1–30.

Carsel RF, Mulkey LA, Lorber MN, Baskin LB (1985) The pesticide root zone model (PRZM): a procedure for evaluating pesticide leaching threats to groundwater. Ecol Modell 30:49–69.

Cheng HH (1990) Pesticides in the Soil Environment: Processes, Impacts, and Modeling. SSSA Book Series 2. Soil Science Society of America, Madison, WI.

Chesters G, Pionke HB, Daniel TC (1974) Extraction and analytical techniques for pesticides in soil, sediment, and water. In: Guenzi WD (ed) Pesticide in Soil and Water. American Society of Agronomy, Madison, WI, pp 451–550.

Chung KH, Ro KS, Ondrus MG (1994) Adsorption and extraction of hydroxyatrazine from wetland sediment. J Environ Sci Health A29:1577–1586.

Clay DE, Holman PW, Clay SA, Schumacher TE, Scholes KA, Bender AR (1994) Agrichemical detection in a shallow unconfined aquifer as influenced by sampling technique. Soil Sci Soc Am J 58:102–104.

Clay SA, Koskinen WC (1990a) Characterization of alachor and atrazine desorption from soils. Weed Sci 38:74–80.

Clay SA, Koskinen WC (1990b) Adsorption and desorption of atrazine, hydroxyatrazine and *s*-glutathione atrazine on two soils. Weed Sci 38:262–266.

Clay SA, Allmaras RR, Koskinen WC, Wyse DL (1988) Desorption of atrazine and cyanazine from soil. J Environ Qual 17:719–723.

Clothier BE, Kirkham MB, Mclean JE (1992) In situ measurements of the effective transport volume for solute moving through soil. Soil Sci Soc Am J 56:733–736.

Dao TW, Lavy TL (1978) Atrazine adsorption on soil as influenced by temperature, moisture content and electrolyte concentration. Weed Sci 26:303–308.

Davidson JM, Rieck CE, Santelman PW (1968) Influence of water flux and porous material on the movement of selected herbicides. Soil Sci Soc Am Proc 32:629–633.

Durand G, Barcelo D (1992) Environmental degradation of atrazine, linuron, and fenitrothion in soil samples. Toxicol Environ Chem 36:225–234.

Edwards WM, Shipitalo MJ, Dick WA, Owens LB (1992a) Rainfall intensity affects transport of water and chemicals through macropores in no-till soil. Soil Sci Soc Am J 56:52–58.

Edwards WM, Shipitalo MJ, Traina SJ, Edwards CA, Owens LB (1992b) Role of *Lumbricus terrestris* (L.) burrows on quality of infiltration water. Soil Biol Biochem 24:1555–1561.

Edwards WM, Shipitalo MJ, Owens LB, Dick WA (1993) Factors affecting preferential flow of water and atrazine through earthworm burrows under continuous no-till corn. J Environ Qual 22:453–457.

Elint M, Arvin E, Jensen BK (1993) Degradation of the pesticides mecoprop and atrazine in unpolluted sandy aquifers. J Environ Qual 22:262–266.

Elling W, Huber SJ, Bankstahl B, Hock B (1987) Atmospheric transport of atrazine: a simple device for its detection. Environ Pollut 48:77–82.

Elrick DE, Erh KT, Krupp HK (1966) Application of miscible displacement techniques to soils. Water Resour Res 2:717–727.

Erickson LE, Lee KH (1989) Degradation of atrazine and related *s*-triazine. CRC Crit Rev Environ Control 19:1–14.

Fleming GF, Wax LM, Simmons FW (1992) Leachability and efficacy of starch-encapsulated atrazine. Weed Technol 6:297–302.

Frank R, Sirons GJ (1985) Dissipation of atrazine residues from soils. Bull Environ Contam Toxicol 34:541–548.

Gaber HM, Inskeep WP, Comfort SD, Wraith JM (1995) Nonequilibrium transport of atrazine through large intact soil cores. Soil Sci Soc Am J 59:60–67.

Gamble DS, Khan SU (1988) Atrazine hydrolysis in aqueous suspensions of humic acid at 25 °C. Can J Chem 66:2605–2617.

Gamble DS, Khan SU (1990) Atrazine in organic soils: chemical speciation during heterogenous catalysis. J Agric Food Chem 38:297–308.

Gamble DS, Khan SU (1992) Atrazine in mineral soils: chemical species and catalyzed hydrolysis. Can J Chem 70:1597–1603.

Gamerdinger AP, Lemley AT, Wagenet RJ (1991) Nonequilibrium sorption and degradation of three 2-chloro-s-triazine herbicides in a soil–water system. J Environ Qual 20:815–822.

Garner WY, Honeycutt RC, Nigg HN (1986) Evaluation of Pesticide in Ground Water. ACS Symposium Series 315. American Chemical Society, Washington, DC.

Gaston LA, Locke MA (1994) Predicting Alachlor mobility using batch sorption kinetic data. Soil Sci 158:345–354.

Gaynor JD, MacTarish DC, Findlay WI (1995) Atrazine and metolachlor loss in surface and subsurface runoff from three tillage treatments in corn. J Environ Qual 24:246–256.

Ghadiri H, Shea PJ, Wicks GA, Haderlie LC (1984) Atrazine dissipation in conventional-till and no-till sorghum. J Environ Qual 13:549–552.

Giardina MC, Giardi MT, Filacchioni G (1980) 4-amino-2-chloro-1,3,5-triazine: a new metabolite of atrazine by a soil bacterium. Agric Biol Chem 44:2067–2072.

Gilchrist GF, Gamble DS, Kodama H, Khan SU (1993) Atrazine interactions with clay minerals: kinetics and equilibria of sorption. J Agric Food Chem 41:1748–1755.

Gish TJ, Helling CS, Mojasevic M (1991) Preferential movement of atrazine and cyanazine under field conditions. Trans ASAE 34:1699–1705.

Gish TJ, Shirmohammadi A, Wienhold BJ (1994) Field-scale mobility and persistence of commercial and starch-encapsulated atrazine and alachlor. J Environ Qual 23:355–359.

Glotfelty DE, Leech MM, Jersey J, Taylor AW (1989) Volatilization and wind erosion of soil surface applied atrazine, simazine, alachlor, and toxaphene. J Agric Food Chem 37:546–551.

Green JD, Horton R, Baker JL (1995) Crop residue effects on the leaching of surface-applied chemicals. J Environ Qual 24:343–351.

Green RE, Yamane VK (1970) Precision in pesticide adsorption measurements. Soil Sci Soc Am Proc 34:353–354.

Green RE, Corey JC (1971) Pesticide adsorption measurement by flow equilibrium and subsequent displacement. Soil Sci Soc Am Proc 35:561–565.

Green RE, Davidson JM, Biggar JW (1980) An assessment of methods for determining adsorption-desorption of organic chemicals. In: Banin A, Kafkafi U (eds) Agrochemicals in Soils. Pergamon Press, New York, pp 73–82.

Green RE, Karickhoff SW (1990) Sorption estimates for modeling. In: Cheng HH (ed) Pesticides in the Soil Environment: Processes, Impacts and Modeling. SSSA Book Series 2. Soil Science Society of America, Madison, WI, pp 79–101.

Grover R, Hance RJ (1970) Effect of ratio of soil to water on adsorption of linuron and atrazine. Soil Sci 109:136–138.

Gschwend PM, Wu SC (1985) On the constancy of sediment-water partition coefficients of hydrophobic organic pollutants. Environ Sci Technol 19:90–96.

Gustafson DI (1989) Groundwater ubiquity score: a simple method for assessing pesticide leachability. Environ Toxicol Chem 8:339–357.

Gustafson DI, Holden LR (1990) Nonlinear pesticide dissipation in soil: a new model based on spatial variability. Environ Sci Technol 24:1032–1038.

Haan CT, Nofziger DL, Ahmed FK (1994) Characterizing chemical transport variability due to natural weather sequences. J Environ Qual 23:349–354.

Hall JK, Hartwig NL, Hoffman LD (1983) Application mode and alternate cropping effects on atrazine losses from a hillside. J Environ Qual 12:336–340.

Hance RJ (1988) Adsorption and bioavailability. In: Grover R (ed) Environmental Chemistry of Herbicides, Vol I. CRC Press, Boca Raton, FL, pp 1–19.

Helling CS, Gish TJ (1986) Soil characteristics affecting pesticide movement into ground water. In: Garner WY, Honeycutt RC, Nigg HN (eds) Evaluation of Pesticides in Ground Water. ACS Symposium Series 315. American Chemical Society, Washington, DC, pp 14–38.

Hiltbold AE, Buchanan GA (1977) Influence of soil pH on persistence of atrazine in the field. Weed Sci 25:515–520.

Huang LQ, Pignatello JJ (1990) Improved extraction of atrazine and metolachlor in field soil samples. J Assoc Anal Chem 73:443–446.

Huang PM, Grover R, McKercher RB (1984) Components and particle size fractions involved in atrazine adsorption by soils. Soil Sci 138:20–24.

Hubbs CW, Lavy TL (1990) Dissipation of norfluranzon and other persistent herbicides in soil. Weed Sci 38:81–88.

Hutson JL, Wagenet RJ (1993) A pragmatic field-scale approach for modeling pesticides. J Environ Qual 22:494–499.

Isensee AR, Helling CS, Gish TJ, Kearney PC, Coffman CB, Zhuang W (1988) Groundwater residues of atrazine, alachlor, and cyanazine under no-tillage practices. Chemosphere 17:165–174.

Jayachandran K, Steinheimer TR, Somasundaram L, Mooreman TB, Kanwar RS, Coats JR (1994) Occurrence of atrazine and degradates as contaminants of subsurface drainage and shallow groundwater. J Environ Qual 23:311–319.

Jaynes DB, Logsdon SD, Horton R (1995) Field method for measuring mobile/immobile water content and solute transfer rate coefficient. Soil Sci Soc Am J 59:352–356.

Johnson DC, Selim HM, Ma L, Southwick LM, Willis GH (1995) Movement of atrazine and nitrate in Sharkey soil: evidence of preferential flow. Agricultural Center Bull. 846. Louisiana State University, Baton Rouge, LA, p 24.

Johnson JA, Farmer WJ (1993) Batch versus column method for determining distribution of organics between soil and water phases. Soil Sci 155:92–99.

Johnson RM, Pepperman AB (1995) Mobility of atrazine from alginate controlled release formulations. J Environ Sci Health B30:27–47.

Jury WA, Spencer WF, Farmer WJ (1983) Use of model for assessing relative volatility, mobility, and persistence of pesticides and other trace organics in soil systems. In: Saxena J (ed) Hazard Assessment of Chemicals: Current Developments, Vol 2. Academic Press, New York, pp 1–43.

Jury WA, Focht DD, Farmer WJ (1987) Evaluation of pesticide groundwater pollution potential from standard indices of soil chemical adsorption and biodegradation. J Environ Qual 16:422–428.

Kalouskova N (1989) Adsorption of atrazine on humic acids. J Environ Sci Health B24:599–617.

Kaufmann DD, Blake J (1970) Degradation of atrazine by soil fungi. Soil Biol Biochem 2:273–280.

Khan SU (1991) Bound (nonextractable) pesticide degradation products in soils. In: Somasundaram L, Coats JR (eds) Pesticide Transformation Products: Fate and Significance in the Environment. ACS Symposium Series 459. American Chemical Society, Washington, DC, pp 108–121.

Kladivko EJ, Van Scoyoc GE, Monke EJ, Oates KM, Pask W (1991) Pesticide and nutrient movement into subsurface tile drains on a silt loam soil in Indiana. J Environ Qual 20:264–270.

Kookana RS, Gerritse RG, Aylmore LAG (1992a) A method for studying nonequilibrium sorption during transport of pesticide in soil. Soil Sci 154:344–349.

Kookana RS, Aylmore LAG, Gerritse RG (1992b) Time-dependent sorption of pesticides during transport in soil. Soil Sci 154:214–225.

Kordel W, Herrchen M, Klein M, Traub-Eberhan U, Kloppel H, Klein W (1992) Determination of the fate of pesticides in outdoor lysimeter experiments. Sci Total Environ 123/124:421–434.

Laird DA, Barriuso E, Dowdy RH, Koskinen WC (1992) Adsorption of atrazine on smectites. Soil Sci Soc Am J 56:62–67.

Laird DA, Yen PY, Koskinen WC, Steinheimer TR, Dowdy RH (1994) Sorption of atrazine on soil clay components. Environ Sci Technol 28:1054–1061.

Leistra M (1980) Transport in solution. In: Hance RJ (ed) Interactions Between Herbicides and the Soil. Academic Press, New York, pp 31–58.

Leonard RA, Kinsel WG, Still DA (1987) GLEAMS: groundwater loading effects of agricultural management system. Trans ASAE 30:1403–1418.

Levanon D (1993) Role of fungi and bacteria in the mineralization of the pesticides atrazine, alachlor, malathion, and carbofuran in soil. Soil Biol Biochem 25: 1097–1105.

Li GC, Felbeck GT Jr (1972) Atrazine hydrolysis as catalyzed by humic acids. Soil Sci 114:201–209.

Libik AW, Romanowski RR (1976) Soil persistence of atrazine and cyanazine. Weed Sci 24:627–629.

Ma L, Southwick LM, Willis GH, Selim HM (1993) Hysteretic characteristics of atrazine adsorption–desorption by a Sharkey soil. Weed Sci 41:627–633.

Ma L, Selim HM (1994a) Predicting atrazine adsorption-desorption in soils: a modified second-order model. Water Resour Res 30:447–456.

Ma L, Selim HM (1994b) Predicting atrazine transport in soils: second-order and multireaction approaches. Water Resour Res 30:3489–3498.

Ma L, Selim HM (1995) Transport of a nonreactive solute in soils: a two-flow domain approach. Soil Sci 159:224–234.

Mandelbaum RT, Wackett LP, Allan DL (1993) Mineralization of the s-triazine ring of atrazine by stable bacterial mixed cultures. Appl Environ Microbiol 59: 1695–1701.

McMahon PB, Chapelle FH, Jagucki ML (1992) Atrazine mineralization potential of alluvial-aquifer sediments under aerobic conditions. Environ Sci Technol 26: 1556–1559.

Mingelgrin U, Gerstl Z (1983) Reevaluation of partitioning as a mechanism of nonionic chemicals adsorption in soils. J Environ Qual 12:1–11.

Muir DCG, Baker BE (1978) The disappearance and movement of three triazine herbicides and several of their degradation products in soil under field conditions. Weed Res 18:111–120.

Nair DR, Schnoor JL (1992) Effect of two electron acceptors and atrazine mineralization rates in soils. Environ Sci Technol 26:2298–2300.

Nash RG (1988) Dissipation from soil. In: Grover R (ed) Environmental Chemistry of Herbicides, Vol I. CRC Press, Boca Raton, FL, pp 131–169.

Nations BK, Hallberg GR (1992) Pesticides in Iowa precipitation. J Environ Qual 21:486–492.

Nicholls PH, Walker A, Baker RJ (1982) Measurement and simulation of the movement and degradation of atrazine and metribuzin in a fallow soil. Pestic Sci 12: 484–494.

Nkedi-Kizza P, Biggar JW, van Genuchten MT, Wierenga PJ, Selim HM, Davidson JM, Nielsen DR (1983) Modeling tritium and chloride-36 transport through an aggregated oxisol. Water Resour Res 19:691–700.

Nkedi-Kizza P, Biggar JM, Selim HM, van Genuchten MT, Wierenga PJ, Davidson JM, Nielsen DR (1984) On the equivalence of two conceptual models for describing ion exchange during transport through an aggregated soil. Water Resour Res 20:1123–1130.

Nofziger DL, Hornsby AG (1986) A microcomputer based management tool for chemical movement in soil. Appl Agric Res 1:50–56.

Pennell KD, Hornsby AG, Jessup RE, Rao PSC (1990) Evaluation of five simulation models for predicating aldicarb and bromide behavior under field conditions. Water Resour Res 26:2679–2693.

Periera WE, Rostad CE (1990) Occurrence, distributions and transport of herbicides and their degradation products in the lower Mississippi river and its tributaries. Environ Sci Technol 24:1400–1406.

Phogat BS, Malik RK, Bhan VM (1984) The rate of atrazine degradation in sterilized and unsterilized soil. Beitr Trop Landwirtsch Veterinarmed 22(H.4):391–396.

Pignatello JJ, Huang LQ (1991) Sorptive reversibility of atrazine and metolachlor residues in field soil samples. J Environ Qual 20:222–228.

Pignatello JJ, Ferrandino FJ, Huang LQ (1993) Elution of aged and freshly added herbicides from a soil. Environ Sci Technol 27:1563–1571.

Pimental D, Levitan L (1986) Pesticides: amounts applied and amounts reaching pests. Bioscience 36:86–91.

Pimental D, Mclaughlin L, Zepp A, Lakitan B (1991) Environmental and economic effects of reducing pesticide use. Bioscience 41:402–409.

Pionke HB, Glotfelty DE, Lucas AD, Urban JB (1988) Pesticide contamination of groundwater in the Mahantango Creek watershed. J Environ Qual 17:76–84.

Radosevich M, Traina SJ, Hao YL, Tuovinen OH (1995) Degradation and mineralization of atrazine by a soil bacterial isolate. Appl Environ Microbiol 61:297–302.

Rao PSC, Davidson JM, Jessup RE, Selim HM (1979) Evaluation of conceptual models for describing non-equilibrium adsorption-desorption of pesticides during steady-flow in soils. Soil Sci Soc Am J 43:22–28.

Rao PSC, Jessup RE, Ralston DE, Davidson JM, Kilcrease DP (1980) Experimental and mathematical description of nonadsorbed solute transfer by diffusion in spherical aggregate. Soil Sci Soc Am J 44:684–688.

Rao PSC, Jessup RE (1983) Sorption and movement of pesticides and other toxic organic substances in soils. In: Chemical Mobility and Reactivity in Soil Systems. SSSA Special Publication 11. Soil Science Society of America, Madison, WI, pp 183–201.

Richter J, Richter O, Marucchini C, Perucci P (1992) Kinetic of degradation of some herbicides in soil samples under controlled conditions. Z Pflanzenernaehr Bodenkd 155:261–267.

Riggle BD, Penner D (1987) Evaluation of pine kraft lignins for controlled release of alachlor and metribuzin. Weed Sci 35:243–246.

Roy WR, Krapac IG (1994) Adsorption and desorption of atrazine and deethylatrazine by low organic carbon geologic materials. J Environ Qual 23:549–556.

Sadeghi AM, Isensee AR (1992) Effect of tillage systems and rainfall patterns on atrazine distribution in soil. J Environ Qual 21:464–469.

Sanchez-Martin MJ, Crisanto T, Arienzo M, Sanchez-Camazano M (1994) Evaluation of the mobility of $^{14}$C-labelled pesticides in soils by thin layer chromatography using a linear analyzer. J Environ Sci Health B29:473–484.

Sauer TJ, Daniel TC (1987) Effect of tillage system on runoff losses of surface-applied pesticides. Soil Sci Soc Am J 51:410–415.

Sauer TJ, Fermanich KJ, Daniel TC (1990) Comparison of the pesticide root zone model simulated and measured pesticide mobility under two tillage systems. J Environ Qual 19:727–734.

Schiavon M (1988a) Studies of the movement and the formation of bound residues of atrazine, of its chlorinated derivations, and of hydroxyatrazine in soil using $^{14}$C-ring-labeled compounds under outdoor conditions. Ecotoxicol Environ Saf 15:55–61.

Schiavon M (1988b) Studies of leaching of atrazine, of its chlorinated derivations, and of hydroxyatrazine in soil using $^{14}$C-ring-labeled compounds under outdoor conditions. Ecotoxicol Environ Saf 15:46–54.

Schocken MJ, Speedie MK (1984) Physiological aspects of atrazine degradation by higher marine fungi. Arch Environ Contam Toxicol 13:707–714.

Schottler SP, Eisenreich SJ, Capel PD (1994) Atrazine, alachlor, and cyanazine in a large agricultural river system. Environ Sci Technol 28:1079–1089.

Schreiber MM, Hickman MV, Vail GD (1993) Starch-encapsulated atrazine: efficacy and transport. J Environ Qual 22:443–453.

Selim HM, Davidson JM, Mansell RS (1976) Evaluation of a two-site adsorption-desorption model for describing solute transport in soils. In: Proceedings, Summer Computer Simulation Conference. Simulation Council, Inc., Washington, DC, pp 444–448.

Selim HM, Amacher MC (1988) A second-order kinetic approach for modeling solute retention and transport in soils. Water Resour Res 24:2061–2075.

Selim HM, Buchter B, Hinz C, Ma L (1992) Modeling the transport and retention of cadmium in soils: multireaction and multicomponent approaches. Soil Sci Soc Am J 56:1004–1015.

Selim HM, Ma L (1995) Transport of reactive solutes in soils: a modified two-region approach. Soil Sci Soc Am J 59:75–82.

Seybold CA, McSweeney K, Lowery B (1994) Atrazine adsorption in sandy soils of Wisconsin. J Environ Qual 23:1291–1297.

Shipitalo MJ, Edwards WM, Dick WA, Owens LB (1990) Initial storm effects on macropore transport of surface applied chemicals in no-till soil. Soil Sci Soc Am J 54:1530–1536.

Sichani SA, Engel BA, Monke EJ, Eigel JD, Kladivko EJ (1991) Validating GLEAMS with pesticide field data on a Clermont silt loam soil. Trans ASAE 34: 1732–1737.

Singh G, Spencer WF, Cliath MM, van Genuchten MT (1990a) Sorption behavior of *s*-triazine and thiocarbamate herbicides on soils. J Environ Qual 19:520–525.

Singh G, Spencer WF, Cliath MM, van Genuchten MT (1990b) Dissipation of *s*-triazine and thiocarbamates from soil as related to soil moisture content. Environ Pollut 66:253–262.

Singh R, Gerritse RG, Aylmore LAG (1989) Adsorption–desorption behavior of selected pesticides in some Western Australian soils. Aust J Soil Res 28:227–243.

Skopp J, Warrick AW (1974) A two-phase model for the miscible displacement of reactive solutes in soils. Soil Sci Soc Am Proc 38:545–550.

Smith AE, Walker A (1989) Prediction of the persistence of the triazine herbicides atrazine, cyanazine, and metribuzine in Regina heavy clay. Can J Soil Sci 69: 587–595.

Smith MC, Thomas DL, Bottcher AB, Campbell KL (1990) Measurement of pesticide transport to shallow ground water. Trans ASAE 33:1573–1582.

Smith MC, Bottcher AB, Campbell KL, Thomas DL (1991) Field testing and comparison of the PRZM and GLEAMS models. Trans ASAE 34:838–847.

Smith WN, Prasher SO, Khan SU, Barthakur NN (1992) Leaching of $^{14}$C-labelled atrazine in long, intact soil columns. Trans ASAE 35:1213–1220.

Somasundaram L, Coats JR (1991) Pesticide Transformation Products: Fate and Significance in the Environment. ACS Symposium Series 459. American Chemical Society, Washington, DC.

Somasundaram L, Coats JR, Racke KD, Shanbhag VM (1991) Mobility of pesticides and their hydrolysis metabolites in soil. Environ Toxicol Chem 10:185–194.

Sorenson BA, Wyse DL, Koskinen WC, Buhler DD, Lueschen WE, Jorgenson MD (1993) Formation and movement of $^{14}$C-atrazine degradation products in a sandy loam soil under field conditions. Weed Sci 41:239–245.

Southwick LM, Willis GH, Bengston RL, Lormand TJ (1990a) Atrazine and Metolachlor in subsurface drain water in Louisiana. J Irrig Drain Eng 116:16–23.

Southwick LM, Willis GH, Bengston RL, Lormand TJ (1990b) Effect of subsurface drainage on runoff losses of atrazine and metolachlor in southern Louisiana. Bull Environ Contam Toxicol 45:113–119.

Southwick LM, Willis GH, Selim HM (1992) Leaching of atrazine from sugarcane in southern Louisiana. J Agric Food Chem 40:1264–1268.

Spalding RF, Snow DD, Cassada DA, Burbach ME (1994) Study of pesticide occurrence in two closely spaced lakes in northeastern Nebraska. J Environ Qual 23: 571–578.

Starr JL, Glotfelty DE (1990) Atrazine and bromide movement through a silt loam soil. J Environ Qual 19:552–558.

Steenhuis TS, Pacenka S, Porter KS (1987) MOUSE: a management model for evaluating groundwater contamination for diffuse surface sources aided by computer graphics. Appl Agric Res 2:277–289.

Stehouwer RC, Dick WA, Traina SJ (1993) Characteristics of earthworm burrow lining affecting atrazine sorption. J Environ Qual 22:181–185.

Stehouwer RC, Dick WA, Traina SJ (1994) Sorption and retention of herbicides in vertically oriented earthworm and artificial burrows. J Environ Qual 23:286–292.

Steinheimer TR (1993) HPLC determination of atrazine and principal degradates in agricultural soils and associated surface and ground water. J Agric Food Chem 41:588–595.

Stevenson FJ (1976) Organic matter reactions involving pesticides in soil. In: Kaufman DD, Still G, Paulson GD, Bandal SK (eds) Bound and Conjugated Pesticide Residues. ACS Symposium Series 29. American Chemical Society, Washington, DC, pp 180–207.

Swanson RA, Dutt GR (1973) Chemistry and physical processes that affect atrazine movement and distribution in soil system. Soil Sci Soc Am Proc 37:872–876.

Thooko LW, Rudra RP, Dickinson WT, Patni NK, Wall GJ (1994) Modeling pesticide transport in subsurface drained soils. Trans ASAE 37:1175–1181.

Thurman EM, Meyer M, Pomes M, Perry CA, Schwab AP (1990) Enzyme-linked immunosorbent assay compared with gas chromatography/mass spectrometry for the determination of triazine herbicide in water. Anal Chem 62:2043–2048.

Topp E, Smith WN, Reynolds WD, Khan SU (1994) Atrazine and metalachlor dissipation in soils incubated in undisturbed cores, repacked cores, and flasks. J Environ Qual 23:693–700.

Travis CC, Etnier EL (1981) A survey of sorption relationships for reactive solutes in soil. J Environ Qual 10:8–17.

Troiano J, Garretson C, Krauter C, Brownell J, Huston J (1993) Influence of amount and method of irrigation water application on leaching of atrazine. J Environ Qual 22:290–298.

U.S. Environmental Protection Agency (USEPA) (1991) National primary drinking water regulations, final rule. Fed Reg 56(20):3526–3594.

Utermann J, Kladivko EJ, Jury WA (1990) Evaluating pesticide migration in tile-drained soils with a transfer function model. J Environ Qual 19:707–714.

van Genuchten MT, Davidson JM, Wierenga PJ (1974) An evaluation of kinetic and equilibrium equations for the prediction of pesticide movement through porous media. Soil Sci Soc Am Proc 38:29–35.

van Genuchten MT, Wierenga PJ (1976) Mass transfer studies in sorbing porous media. I. Analytical solution. Soil Sci Soc Am J 40:473–480.

van Genuchten MT, Wierenga PJ (1977) Mass transfer studies in sorbing porous media. II. Experimental evaluation with tritium ($^3H_2O$). Soil Sci Soc Am J 41: 272–277.

van Genuchten MT, Wierenga PJ, O'Conner GA (1977) Mass transfer studies in sorbing porous media: III. Experimental evaluations with 2,4,5-T. Soil Sci Soc Am J 41:278–285.

van Genuchten MT (1981) Non-equilibrium transport parameters from miscible displacement experiments. Research Report 119, U.S. Salinity Lab, Riverside, CA.

van Genuchten MT, Dalton FN (1986) Models for simulating salt movement in aggregated field soils. Geoderma 38:165–183.

van Genuchten MT, Wagenet RJ (1989) Two-site/two-region models for pesticide transport and degradation: theoretical development and analytical solutions. Soil Sci Soc Am J 53:1303–1310.

Von Stryk FG, Bolton EF (1977) Atrazine residues in tile-drain water from corn plots as affected by cropping practices and fertility levels. Can J Soil Sci 57:249–253.

Wagenet RJ, Hutson JL (1986) Predicting the fate of nonvolatile pesticides in the unsaturated zone. J Environ Qual 15:315–322.

Wang Z, Gamble DS, Langford CH (1990) Interaction of atrazine with laurentian fulvic acid: binding and hydrolysis. Anal Chim Acta 232:181–188.

Wang Z, Gamble DS, Langford CH (1992) Interaction of atrazine with Laurentian soil. Environ Sci Technol 26:560–565.

Wauchope RD, Myers RS (1985) Adsorption-desorption kinetics of atrazine and linuron in freshwater-sediment aqueous slurries. J Environ Qual 14:132–136.

Wauchope RD, Buttler TM, Hornsby AG, Augustjin-Beckers PWM, Burt HM (1992) The SCS/ARS/CES pesticide properties database for environmental decision making. Rev Environ Contam Toxicol 123:1–164.

Weber JB (1993) Ionization and sorption of fomesafen and atrazine by soils and soil constituents. Pestic Sci 39:31–38.

Weed DAJ, Kanwar RS, Stoltenberg DE, Pfeiffer RL (1995) Dissipation and distribution of herbicides in the soil profile. J Environ Qual 24:68–79.

Welhouse GJ, Bleam WF (1993) Cooperative hydrogen bonding of atrazine. Environ Sci Technol 27:500–505.

Whang JM, Schomburg CJ, Glotfelty DE, Taylor AW (1993) Volatilization of fonofos, chlorpyrifos, and atrazine from conventional and no-till surface soils in the field. J Environ Qual 22:173–180.

White AW, Barnett AP, Wright BG, Holladay JH (1967) Atrazine losses from fallow land caused by runoff and erosion. Environ Sci Technol 1:740–744.

White JL (1976) Clay-pesticide interactions. In: Kaufman DD, Still G, Paulson GD, Bandal SK (eds) Bound and Conjugated Pesticide Residues. ACS Symposium Series 29. American Chemical Society, Washington, DC, pp 208–218.

Widmer SK, Olson JM, Koskinen WC (1993) Kinetics of atrazine hydrolysis in water. J Environ Sci Health B28:19–28.

Wienhold BJ, Gish TJ (1992) Effect of water potential, temperature, and soil microbial activity on release of starch-encapsulated atrazine and alachlor. J Environ Qual 21:382–386.

Wienhold BJ, Sadeghi AM, Gish TJ (1993) Effect of starch encapsulation and temperature on volatilization of atrazine and alachlor. J Environ Qual 22:162–166.

Wienhold BJ, Gish TJ (1994) Effect of formulation and tillage practice on volatilization of atrazine and alachlor. J Environ Qual 23:292–298.

Wisconsin Department of Natural Resources (WDNR) (1988) Summary of groundwater pesticides monitoring for 07/01/83 through 06/30/87. Prepared by staff of WDNR, June 23, 1988.

Yaron B, Gerstl Z, Spencer WF (1985) Behavior of herbicides in irrigated soils. Adv Soil Sci 3:121–211.

---

Manuscript received March 16, 1995; accepted May 12, 1995.

# Index